BEING YOU

BEING YOU

A New Science of Consciousness

ANIL SETH

DUTTON

DUTTON

An imprint of Penguin Random House LLC
penguinrandomhouse.com

LIBRARY OF CONGRESS CATALOGING-IN-PUBLICATION DATA
has been applied for.

ISBN 9781524742874 (hardcover)
ISBN 9781524742881 (ebook)

Printed in the United States of America

1 3 5 7 9 10 8 6 4 2

BOOK DESIGN BY ELLEN CIPRIANO

The Brain—is wider than the Sky—
For—put them side by side—
The one the other will contain
With ease—and You—beside—
EMILY DICKINSON

For my mother,
Ann Seth, and in memory of
my father, Bhola Nath Seth

CONTENTS

IV: OTHER

PROLOGUE

FIVE YEARS AGO, for the third time in my life, I ceased to exist. I was having a small operation and my brain was filling with anesthetic. I remember sensations of blackness, detachment, and falling apart . . .

General anesthesia is very different from going to sleep. It has to be; if you were asleep, the surgeon's knife would quickly wake you up. States of deep anesthesia have more in common with catastrophic conditions like coma and the vegetative state, where consciousness is completely absent. Under profound anesthesia, the brain's electrical activity is almost entirely quieted—something that never happens in normal life, awake or asleep. It is one of the miracles of modern medicine that anesthesiologists can routinely alter people's brains so that they enter and return from such deeply unconscious states. It's an act of transformation, a kind of magic: anesthesia is the art of turning people into objects.

The objects, of course, get turned back into people. So I returned, drowsy and disoriented but definitely *there*. No time seemed to have passed. Waking from a deep sleep, I am sometimes confused about the time, but there is always the impression that at least some amount of time has gone by, of a continuity between my consciousness then and my consciousness now. Under general anesthesia, things are different. I could have been under for five minutes, five hours, five years—or even fifty. And "under" doesn't quite express it. I was simply not there, a premonition of the total oblivion of death, and, in its absence of anything, a strangely comforting one.

General anesthesia doesn't just work on your brain, or on your mind. It works on your consciousness. By altering the delicate electrochemical balance within the neural circuitry inside your head, the basic ground state of what it is to "be" is—temporarily—abolished. In this process lies one of the greatest remaining mysteries in science, and in philosophy too.

Somehow, within each of our brains, the combined activity of billions of neurons, each one a tiny biological machine, is giving rise to a conscious experience. And not just any conscious experience, *your* conscious experience, right here, right now. *How does this happen? Why do we experience life in the first person?*

I have a childhood memory of looking in the bathroom mirror, and for the first time realizing that my experience at that precise moment—the experience of *being me*—would at some point come to an end, and that "I" would die. I must have been about eight or nine years old, and like all early memories, it is unreliable. But perhaps it was at this moment that I also realized that if my consciousness could end, then it must depend in some way on the stuff I was made of—on the physical materiality of my body and my brain. It seems to me that I've been grappling with this mystery, in one way or another, ever since.

As an undergraduate student at Cambridge University in the early nineties, a teenage romance with physics and philosophy broadened into a fascination with psychology and neuroscience, even though at the time these fields seemed to avoid, even outlaw, all mention of consciousness. My PhD research took me on a long and unexpectedly valuable detour through artificial intelligence and robotics, before a six-year stint at the Neurosciences Institute in San Diego, on the shores of the Pacific, finally delivered the chance to investigate the brain basis of consciousness directly. There, I worked with the Nobel Laureate Gerald Edelman—one of the most significant figures in bringing consciousness back into view as a legitimate scientific focus.

Now, for more than a decade, I've been co-director of a research center—the Sackler Centre for Consciousness Science at the University of Sussex—nestled among the gentle green hills of the South Downs by the seaside city of Brighton. Our center brings together neuroscientists, psychologists, psychiatrists, brain imagers, virtual-reality wizards and mathematicians, and philosophers, all of us trying to open new windows onto the brain basis of conscious experience.

WHETHER YOU'RE A SCIENTIST OR NOT, consciousness is a mystery that matters. For each of us, our conscious experience is *all there is*. Without it, there is nothing at all: no world, no self, no interior and no exterior.

Imagine that a future version of me, perhaps not so far away, offers you the deal of a lifetime. I can replace your brain with a machine that is its equal in every way, so that from the outside, nobody could tell the difference. This new machine has many advantages— it is immune to decay, and perhaps it will allow you to live forever.

But there's a catch. Since even future-me is not sure how real brains give rise to consciousness, I can't guarantee that you will have any conscious experiences at all, should you take up this offer.

Maybe you will, if consciousness depends only on functional capacity, on the power and complexity of the brain's circuitry, but maybe you won't, if consciousness depends on a specific biological material—neurons, for example. Of course, since your machine-brain leads to identical behavior in every way, when I ask new-you whether you are conscious, new-you will say yes. But what if, despite this answer, life—for *you*—is no longer in the first person?

I suspect you wouldn't take the deal. Without consciousness, it may hardly matter whether you live for another five years or another five hundred. In all that time, *there would be nothing it would be like to be you.*

Philosophical games aside, the practical importance of understanding the brain basis of consciousness is easy to appreciate. General anesthesia has to count as one of the greatest inventions of all time. Less happily, distressing disturbances of consciousness can accompany brain injuries and mental illnesses for the increasing number of us, me included, who encounter these conditions. And for each one of us, conscious experiences change throughout life, from the blooming and buzzing confusion of early life, through the apparent though probably illusory and certainly not universal clarity of adulthood, and on to our final drift into the gradual—and for some, disorientingly rapid—dissolution of the self as neurodegenerative decay sets in. At each stage in this process, you exist, but the notion that there is a single unique conscious self (a soul?) that persists over time may be grossly mistaken. Indeed, one of the most compelling aspects of the mystery of consciousness is the nature of *self*. Is consciousness possible without self-consciousness, and if so, would it still matter so much?

Answers to difficult questions like these have many implications for how we think about the world and the life it contains. When does consciousness begin in development? Does it emerge at birth, or is it present even in the womb? What about consciousness in nonhuman animals—and not just in primates and other mammals, but in otherworldly creatures like the octopus and perhaps even in simple organisms such as nematode worms or bacteria? Is there anything it is like to be an *Escherichia coli*, or a sea bass? What about future machines? Here, we ought to be concerned not just about the power that new forms of artificial intelligence are gaining over us, but also about whether and when *we* need to take an ethical stance toward *them*. For me, these questions evoke the uncanny sympathy I felt when watching Dave Bowman destroy HAL's personality in the film *2001 A Space Odyssey*, by the simple act of removing its memory banks, one by one. In the greater empathy elicited by the plight of Ridley Scott's replicants in *Blade Runner*, there is a clue about the importance of our nature as *living machines* for the experience of being a conscious self.

THIS BOOK IS ABOUT the neuroscience of consciousness: the attempt to understand how the inner universe of subjective experience relates to, and can be explained in terms of, biological and physical processes unfolding in brains and bodies. This is the project that has captivated me throughout my career, and I believe it has now reached a point at which glimmerings of answers are beginning to emerge.

These glimmerings already change, and change dramatically, how we think about conscious experiences of the world around us, and of ourselves within it. The way we think about consciousness touches every aspect of our lives. A science of consciousness is nothing less than an account of who we are, of what it is to be me, or to be you, and of why there is anything it is like to "be" at all.

The story I will tell is a personal view, shaped over many years of research, contemplation, and conversation. The way I see it, consciousness won't be "solved" in the same way that the human genome was decoded or the reality of climate change established. Nor will its mysteries suddenly yield to a single eureka-like insight—a pleasant but usually inaccurate myth about how scientific understanding progresses.

For me, a science of consciousness should explain how the various properties of consciousness depend on, and relate to, the operations of the neuronal wetware inside our heads. The goal of consciousness science should not be—at least not primarily—to explain why consciousness happens to be part of the universe in the first place. Nor should it be to understand how the brain works in all its complexity, while sweeping the mystery of consciousness away under the carpet. What I hope to show you is that by accounting for properties of consciousness, in terms of mechanisms in brains and bodies, the deep metaphysical whys and hows of consciousness become, little by little, less mysterious.

I use the word "wetware" to underline that brains are not computers made of meat. They are chemical machines as much as they are electrical networks. Every brain that has ever existed has been part of a living body, embedded in and interacting with its environment—an environment which in many cases contains other embodied brains. Explaining the properties of consciousness in terms of biophysical mechanisms requires understanding brains—and conscious minds—as *embodied* and *embedded* systems.

In the end, I want to leave you with a new conception of the self—that aspect of consciousness which for each of us is probably the most meaningful. An influential tradition, dating back at least as far as Descartes in the seventeenth century, held that nonhuman animals lacked conscious selfhood because they did not have rational minds to guide their behavior. They were "beast machines": flesh automatons without the ability to reflect on their own existence.

I don't agree. In my view, consciousness has more to do with being alive than with being intelligent. We are conscious selves precisely *because* we are beast machines. I will make the case that the experiences of *being you*, or of *being me*, emerge from the way the brain predicts and controls the internal state of the body. The essence of selfhood is neither a rational mind nor an immaterial soul. It is a deeply embodied biological *process*, a process that underpins the simple feeling of being alive that is the basis for all our experiences of self, indeed for any conscious experience at all. *Being you* is literally about your body.

This book is divided into four parts. In the first part, I explain my approach to the scientific study of consciousness. This part also deals with the question of conscious *level*—of how conscious someone or something can be—and with progress in attempts to "measure" consciousness. The second part takes on the topic of conscious *content*—of what you are conscious of, when you are conscious. Part three turns the focus inward, to the *self*, and to all the varied experiences that conscious selfhood entails. The fourth and final part—*other*—explores what this new way of understanding consciousness can say about other animals, and about the possibility of sentient machines. By the end of the book, you'll understand that our conscious experiences of the world and the self are forms of brain-based prediction—"controlled hallucinations"—that arise with, through, and *because of* our living bodies.

DESPITE HIS TARNISHED REPUTATION among neuroscientists, Sigmund Freud was right about many things. Looking back through the history of science, he identified three "strikes" against the perceived self-importance of the human species, each marking a major scientific advance that was strongly resisted at the time. The first was by Copernicus, who showed with his heliocentric theory that the Earth rotates around the sun, and not the other way around. With this dawned the realization that we are not at the center of the universe; we are just a speck somewhere out there in the vastness, a pale blue dot suspended in the abyss. Next came Darwin, who revealed that we share common ancestry with all other living things, a realization that is—astonishingly—still resisted in some parts of the world even today. Immodestly, Freud's third strike against human exceptionalism was his own theory of the unconscious mind, which challenged the idea that our mental lives are under our conscious, rational control. While he may have been off target in the details, Freud was absolutely right to point out that a naturalistic explanation of mind and consciousness would be a further, and perhaps final, dethronement of humankind.

These shifts in how we see ourselves are to be welcomed. With each new advance in our understanding comes a new sense of wonder, and a new ability to see ourselves as less *apart from*, and more *a part of*, the rest of nature.

Our conscious experiences are part of nature just as our bodies are, just as our world is. And when life ends, consciousness will end too. When I think about this, I am transported back to my experience—my *non*-experience—of anesthesia. To its oblivion, perhaps comforting, but oblivion nonetheless. The novelist Julian Barnes, in his meditation on mortality, puts it perfectly. When the end of consciousness comes, there is nothing—really *nothing*—to be frightened of.

I

LEVEL

1

THE REAL PROBLEM

WHAT IS CONSCIOUSNESS?

For a conscious creature, there is something that it is like to *be* that creature. There is something it is like to be me, something it is like to be you, and probably something it is like to be a sheep, or a dolphin. For each of these creatures, subjective experiences are happening. It *feels like something* to be me. But there is almost certainly nothing it is like to be a bacterium, a blade of grass, or a toy robot. For these things, there is (presumably) never any subjective experience going on: no inner universe, no awareness, no consciousness.

This way of putting things is most closely associated with the philosopher Thomas Nagel, who in 1974 published a now legendary article called "What is it like to be a bat?" in which he argued that while we humans could never experience the experiences of a bat, there nonetheless would be something it is like for the bat, to be a bat.* I've always favored Nagel's approach because it emphasizes *phenomenology*: the subjective properties of conscious experience,

---

* This paper is one of the most influential in all philosophy of mind. According to Nagel, "an organism has conscious mental states if and only if there is something that it is like to *be* that organism—something it is like *for* the organism." Nagel (1974), p. 2 (italics in original).

such as why a visual experience has the form, structure, and qualities that it does, as compared to the subjective properties of an emotional experience, or of an olfactory experience. In philosophy, these
properties are sometimes also called "qualia": the redness of red, the
pang of jealousy, the sharp pain or dull throb of a toothache.

For an organism to be conscious, it has to have some kind of
phenomenology for itself. *Any* kind of experience—*any* phenomenological property—counts as much as any other. Wherever there is
experience, there is phenomenology; and wherever there is phenomenology, there is consciousness. A creature that comes into
being only for a moment will be conscious just as long as there is
something it is like to be it, even if all that's happening is a fleeting
feeling of pain or pleasure.

We can usefully distinguish the phenomenological properties of
consciousness from its *functional* and *behavioral* properties. These
refer to the roles that consciousness may play in the operations of
our minds and brains, and to the behaviors an organism is capable
of, by virtue of having conscious experiences. Although the functions and behaviors associated with consciousness are important
topics, they are not the best places to look for definitions. Consciousness is first and foremost about subjective experience—it is
about phenomenology.

This may seem obvious, but it wasn't always so. At various times
in the past, being conscious has been confused with having language,
being intelligent, or exhibiting behavior of a particular kind. But consciousness does not depend on outward behavior, as is clear during
dreaming and for people suffering states of total bodily paralysis.
To hold that language is needed for consciousness would be to say
that babies, adults who have lost language abilities, and most if not
all nonhuman animals lack consciousness. And complex abstract

thinking is just one small part—though possibly a distinctively human part—of being conscious.

Some prominent theories in the science of consciousness continue to emphasize function and behavior over phenomenology. Foremost among these is the "global workspace" theory, which has been developed over many years by the psychologist Bernard Baars and the neuroscientist Stanislas Dehaene, among others. According to this theory, mental content (perceptions, thoughts, emotions, and so on) becomes conscious when it gains access to a "workspace," which—anatomically speaking—is distributed across the frontal and parietal regions of the cortex. (The cerebral cortex is the massively folded outer surface of the brain, made up of tightly packed neurons.*) When mental content is broadcast within this cortical workspace, we are conscious of it, and it can be used to guide behavior in much more flexible ways than is the case for unconscious perception. For example, I am consciously aware of a glass of water on the table in front of me. I could pick it up and drink it, throw it over my computer (tempting), write a poem about it, or take it back into the kitchen now that I realize it's been there for days. Unconscious perception does not allow this degree of behavioral flexibility.

Another prominent theory, called "higher-order thought" theory, proposes that mental content becomes conscious when there is a "higher-level" cognitive process that is somehow oriented toward it, rendering it conscious. In this theory, consciousness is closely tied to processes like *metacognition*—meaning "cognition about cognition"—which again emphasizes functional properties over

---

* Each hemisphere of the cerebral cortex has four lobes. The frontal lobes are at the front. The parietal lobes are toward the back and off to the sides. The occipital lobes are at the back, and the temporal lobes are at the sides, near the ears. Some people identify a fifth lobe—the limbic lobe—deep inside the brain.

phenomenology (though less so than global workspace theory). Like global workspace theory, higher-order thought theories also emphasize frontal brain regions as key for consciousness.

Although these theories are interesting and influential, I won't have much more to say about either in this book. This is because they both foreground the functional and behavioral aspects of consciousness, whereas the approach I will take starts from phenomenology—from experience itself—and only from there has things to say about function and behavior.

The definition of consciousness as "any kind of subjective experience whatsoever" is admittedly simple and may even sound trivial, but this is a good thing. When a complex phenomenon is incompletely understood, prematurely precise definitions can be constraining and even misleading. The history of science has demonstrated many times over that useful definitions evolve in tandem with scientific understanding, serving as scaffolds for scientific progress, rather than as starting points, or ends in themselves. In genetics, for example, the definition of a "gene" has changed considerably as molecular biology has advanced. In the same way, as our understanding of consciousness develops, its definition—or definitions—will evolve too. If, for now, we accept that consciousness is first and foremost about phenomenology, then we can move on to the next question.

How DOES CONSCIOUSNESS HAPPEN? How do conscious experiences relate to the biophysical machinery inside our brains and our bodies? How indeed do they relate to the swirl of atoms or quarks or superstrings, or to whatever it is that the entirety of our universe ultimately consists in?

The classic formulation of this question is known as the "hard problem" of consciousness. This expression was coined by the Australian philosopher David Chalmers in the early 1990s and it has set the agenda for much of consciousness science ever since. Here is how he describes it:

> It is undeniable that some organisms are subjects of experience. But the question of how it is that these systems are subjects of experience is perplexing. Why is it that when our cognitive systems engage in visual and auditory information-processing, we have visual or auditory experience: the quality of deep blue, the sensation of middle C? How can we explain why there is something it is like to entertain a mental image, or to experience an emotion? It is widely agreed that experience arises from a physical basis, but we have no good explanation of why and how it so arises. Why should physical processing give rise to a rich inner life at all? It seems objectively unreasonable that it should, and yet it does.

Chalmers contrasts this hard problem of consciousness with the so-called easy problem—or easy problems—which have to do with explaining how physical systems, like brains, can give rise to any number of functional and behavioral properties. These functional properties include things like processing sensory signals, selection of actions and the control of behavior, paying attention, the generation of language, and so on. The easy problems cover all the things that beings like us can do and that can be specified in terms of a function—how an input is transformed into an output—or in terms of a behavior.

Of course, the easy problems are not easy at all. Solving them will occupy neuroscientists for decades or centuries to come. Chalmers's point is that the easy problems are easy to solve in principle, while the same cannot be said for the hard problem. More precisely, for Chalmers, there is no conceptual obstacle to easy problems eventually yielding to explanations in terms of physical mechanisms. By contrast, for the hard problem it seems as though no such explanation could ever be up to the job. (A "mechanism"—to be clear—can be defined as a system of causally interacting parts that produce effects.) Even after all the easy problems have been ticked off, one by one, the hard problem will remain untouched. "[E]ven when we have explained the performance of all the functions in the vicinity of experience—perceptual discrimination, categorization, internal access, verbal report—there may still remain a further unanswered question: *Why is the performance of these functions accompanied by experience?*"

The roots of the hard problem extend back to ancient Greece, perhaps even earlier, but they are particularly visible in René Descartes's seventeenth-century sundering of the universe into mind stuff, *res cogitans*, and matter stuff, *res extensa*. This distinction inaugurated the philosophy of dualism, and has made all discussions of consciousness complicated and confusing ever since. This confusion is most evident in the proliferation of different philosophical frameworks for thinking about consciousness.

Take a deep breath, here come the "isms."

My preferred philosophical position, and the default assumption of many neuroscientists, is *physicalism*. This is the idea that the universe is made of physical stuff, and that conscious states are either identical to, or somehow emerge from, particular arrangements of this physical stuff. Some philosophers use the term *materialism* instead of physicalism, but for our purposes they can be treated synonymously.

At the other extreme to physicalism is *idealism*. This is the idea—often associated with the eighteenth-century bishop George Berkeley—that consciousness or mind is the ultimate source of reality, not physical stuff or matter. The problem isn't how mind emerges from matter, but how matter emerges from mind.

Sitting awkwardly in the middle, *dualists* like Descartes believe that consciousness (mind) and physical matter are separate substances or modes of existence, raising the tricky problem of how they ever interact. Nowadays, few philosophers or scientists would explicitly sign up for this view. But for many people, at least in the West, dualism remains beguiling. The seductive intuition that conscious experiences *seem* nonphysical encourages a "naïve dualism" where this "seeming" drives beliefs about how things actually are. As we'll see throughout this book, the way things seem is often a poor guide to how they actually are.

One particularly influential flavor of physicalism is *functionalism*. Like physicalism, functionalism is a common and often unstated assumption of many neuroscientists. Many who take physicalism for granted also take functionalism for granted. My own view, however, is to be agnostic and slightly suspicious.

Functionalism is the idea that consciousness does not depend on what a system is made of (its physical constitution), but only on what the system does, on the functions it performs, on how it transforms inputs into outputs. The intuition driving functionalism is that mind and consciousness are forms of information processing which can be implemented by brains, but for which biological brains are not strictly necessary.

Notice how the term "information processing" sneaked in here unannounced (as it also did in the quote from Chalmers a few pages back). This term is so prevalent in discussions of mind, brain, and consciousness that it's easy to let it slide by. This would be a mistake, because the suggestion that the brain "processes information" conceals some strong assumptions. Depending on who's doing the assuming, these range from the idea that the brain is some kind of computer, with mind (and consciousness) being the software (or "mindware"), to assumptions about what information itself actually *is*. All of these assumptions are dangerous. Brains are very different from computers, at least from the sorts of computers that we are familiar with. And the question of what information "is" is almost as vexing as the question of what consciousness is, as we'll see later on in this book. These worries are why I'm suspicious of functionalism.

Taking functionalism at face value, as many do, carries the striking implication that consciousness is something that can be *simulated* on a computer. Remember that for functionalists, consciousness depends only on what a system does, not on what it is made of. This means that if you get the functional relations right— if you ensure that a system has the right kind of "input-output mappings"—then this will be enough to give rise to consciousness. In other words, for functionalists, *simulation* means *instantiation*— it means coming into being, in reality.

How reasonable is this? For some things, simulation certainly counts as instantiation. A computer that plays Go, such as the world-beating AlphaGo Zero from the British artificial intelligence company DeepMind, *is actually playing Go*. But there are many situations where this is not the case. Think about weather forecasting. Computer simulations of weather systems, however detailed they may be, do not get wet or windy. Is consciousness more like Go or more like the weather? Don't expect an answer—there isn't one, at least not yet. It's enough to appreciate that there's a valid question here. This is why I'm agnostic about functionalism.

There are two more "isms," then we're done.

The first is *panpsychism*. Panpsychism is the idea that consciousness is a fundamental property of the universe, alongside other fundamental properties such as mass/energy and charge; that it is present to some degree everywhere and in everything. People sometimes make fun of panpsychism for claiming things like stones and spoons are conscious in the same sort of way that you or I are, but these are usually deliberate misconstruals designed to make it look silly. There are more sophisticated versions of the idea, some of which we will meet in later chapters, but the main problems with panpsychism don't lie with its apparent craziness—after all, some crazy ideas turn out to be true, or at least useful. The main problems are that it doesn't really explain anything and that it doesn't lead to testable hypotheses. It's an easy get-out to the apparent mystery posed by the hard problem, and taking it on ushers the science of consciousness down an empirical dead end.

Finally, there's *mysterianism*, which is associated with the philosopher Colin McGinn. Mysterianism is the idea that there may exist a complete physical explanation of consciousness—a full solution to Chalmers's hard problem—but that we humans just aren't clever enough, and never will be clever enough, to discover this solution, or even to recognize a solution if it were presented to us by super-smart aliens. A physical understanding of consciousness exists, but it lies as far beyond us as an understanding of cryptocurrency lies beyond frogs. It is cognitively closed to us by our species-specific mental limitations.

What can be said about mysterianism? There may well be things we will never understand, thanks to the limitations of our brains and minds. Already, no single person is able to fully comprehend how an Airbus A380 works. (And yet I'm happy to sit in one, as I did one time on the way home from Dubai.) There are certainly things which remain cognitively inaccessible to most of us, even if they are understandable by humans in principle, like the finer points of string theory in physics. Since brains are physical systems with finite resources, and since some brains seem incapable of understanding some things, it seems inescapable that there must be some things which are the case, but which no human could ever understand. However, it is unjustifiably pessimistic to preemptively include consciousness within this uncharted domain of species-specific ignorance.

One of the more beautiful things about the scientific method is that it is cumulative and incremental. Today, many of us can understand things that would have seemed entirely incomprehensible *even in principle* to our ancestors, maybe even to scientists and philosophers working just a few decades ago. Over time, mystery after mystery has yielded to the systematic application of reason and experiment. If we take mysterianism as a serious option we might as well all give up and go home. So, let's not.

These "isms" provide different ways of thinking about the relationship between consciousness and the universe as a whole. When weighing their merits and demerits, it's important to recognize that what matters most is not which framework is "right" in the sense of being provably true, but which is most useful for advancing our understanding of consciousness. This is why I tend toward a functionally agnostic flavor of physicalism. To me, this is the most pragmatic and productive mindset to adopt when pursuing a science of consciousness. It is also, as far as I am concerned, the most intellectually honest.

DESPITE ITS APPEAL, physicalism is by no means universally accepted among consciousness researchers. One of the most common challenges to physicalism is the so-called "zombie" thought experiment. The zombies in question here are not the brain-munching semi-corpses from the movies—our zombies are "philosophical zombies." But we need to get rid of them all the same, since otherwise the prospect of a natural, physicalist explanation of consciousness is dead in the water before we get started.

A philosophical zombie is a creature that is indistinguishable from a conscious creature, but which lacks consciousness. A zombie Anil Seth would look like me, act like me, walk like me, and talk like me, but there would be nothing it is like to *be* it, no inner universe, no felt experience. Ask zombie Anil if he is conscious, and he will say, "Yes, I'm conscious." Zombie Anil would even have written various essays on the neuroscience of consciousness, including some thoughts about the questionable relevance of philosophical zombies to this topic. But none of this would involve any conscious experience whatsoever.

Here's why the zombie idea is supposed to provide an argument against physicalist explanations of consciousness. If you can imagine a zombie, this means you can conceive of a world that is indistinguishable from our world, but in which no consciousness is happening. And if you can conceive of such a world, then consciousness cannot be a physical phenomenon.

And here's why it doesn't work. The zombie argument, like

many thought experiments that take aim at physicalism, is a conceivability argument, and conceivability arguments are intrinsically weak. Like many such arguments, it has a plausibility that is inversely related to the amount of knowledge one has.

Can you imagine an A380 flying backward? Of course you can. Just imagine a large plane in the air, moving backward. Is such a scenario *really* conceivable? Well, the more you know about aerodynamics and aeronautical engineering, the less conceivable it becomes. In this case, even a minimal knowledge of these topics makes it clear that planes cannot fly backward. It just *cannot be done.**

It's the same with zombies. In one sense it's trivial to imagine a philosophical zombie. I just picture a version of myself wandering around without having any conscious experiences. But can I *really* conceive this? What I'm being asked to do, *really*, is to consider the capabilities and limitations of a vast network of many billions of neurons and gazillions of synapses (the connections between neurons), not to mention glial cells and neurotransmitter gradients and other neurobiological goodies, all wrapped into a body interacting with a world which includes other brains in other bodies. Can I do this? Can anyone do this? I doubt it. Just as with the A380, the more one knows about the brain and its relation to conscious experiences and behavior, the less conceivable a zombie becomes.†

Whether something is conceivable or not is often a psychological observation about the person doing the conceiving, not an

---

* A helicopter, which can fly backward, is not a plane. I was oddly happy to discover that the origin of the word "helicopter" is not a combination of "heli" and "copter," as I'd always assumed, but rather "helico" (spiral) and "pter" (wing). They make much more sense now.

† The adult human brain contains an estimated 86 billion neurons, and about a thousandfold more connections. If you counted one connection every second, it would take you nearly 3 million years to finish. What's more, it's increasingly apparent that even single neurons are capable of carrying out highly complex functions all by themselves.

insight into the nature of reality. This is the weakness of zombies. We are asked to imagine the unimaginable, and through this act of illusory comprehension, conclusions are drawn about the limits of physicalist explanation.

WE'RE NOW READY TO MEET what I call the *real problem* of consciousness. This is a way of thinking about consciousness science that has taken shape for me over many years, assimilating and building on the insights of many others. Addressing the real problem is, I believe, the approach by which a science of consciousness is most likely to succeed.

According to the real problem, the primary goals of consciousness science are to *explain*, *predict*, and *control* the phenomenological properties of conscious experience. This means explaining why a particular conscious experience is the way it is—why it has the phenomenological properties that it has—in terms of physical mechanisms and processes in the brain and body. These explanations should enable us to predict when specific subjective experiences will occur, and enable their control through intervening in the underlying mechanisms. In short, addressing the real problem requires explaining *why* a particular pattern of brain activity—or other physical process—maps to a particular kind of conscious experience, not merely establishing that it does.

The real problem is distinct from the hard problem, because it is not—at least not in the first instance—about explaining why and how consciousness is part of the universe in the first place. It does not hunt for a special sauce that can magic consciousness from mere mechanism (or the other way around). It is also distinct from the easy problem(s), because it focuses on phenomenology rather than on function or behavior. It doesn't sweep the subjective aspects of consciousness away under the carpet. And because of its emphasis on mechanisms and processes, the real problem aligns naturally with a physicalist worldview on the relationship between matter and mind.

To clarify these distinctions, let's ask how the different approaches might attempt to explain the subjective experience of "redness."

From an easy problem perspective, the challenge is to explain all the mechanistic, functional, and behavioral properties associated with experiencing redness: how specific wavelengths of light activate the visual system, the conditions under which we say things like "that object is red," typical behavior at traffic lights, how red things sometimes induce emotional responses of a particular kind, and so on.

Left untouched by the easy problem approach, by design, is any explanation of why and how these functional, mechanistic, and behavioral properties should be accompanied by any phenomenology whatsoever—in this case, the phenomenology of "redness." The existence of subjective experience, as opposed to no experience, is the dominion of the hard problem. No matter how much mechanistic information you're given, it will never be unreasonable for you to ask, "Fine, but why is this mechanism associated with conscious experience?" If you take the hard problem to heart, you will always suspect an explanatory gap between mechanistic explanations and the subjective experience of "seeing red."

The real problem accepts that conscious experiences exist and focuses primarily on their phenomenological properties. For example: experiences of redness are visual, they usually but not always attach to objects, they seem to be properties of surfaces, they have different levels of saturation, they define a category among other color experiences though they can smoothly vary within that category, and so on. Importantly, these are all properties *of the experience itself*, not—at least not primarily—of the functional properties or behaviors associated with that experience. The challenge for the real problem is to explain, predict, and control these phenomenological properties, in terms of things happening in the brain and body. We would like to know what it is about specific patterns of activity in the brain—such as the complex looping activity in the visual cortex*—that explains (and predicts, and controls) why an experience, such as the experience of redness, is the particular way it is, and not some other way. Why it is not like blueness, or toothache, or jealousy.

Explanation, prediction, and control. These are the criteria by which most other scientific projects are assessed, regardless of how mystifying their target phenomena might initially appear. Physicists have made enormous strides in unraveling the secrets of the universe—in explaining, predicting, and controlling its properties—but are still flummoxed when it comes to figuring out what the universe is made of or why it exists. In just the same way, consciousness science can make great progress in shedding light on the properties and nature of conscious experiences without it being necessary to explain how or why they happen to be part of the universe in which we live.

Nor should we necessarily expect scientific explanations to always be intuitively satisfying. In physics, quantum mechanics is notoriously counterintuitive but is nonetheless widely accepted as providing our current best grip on the nature of physical reality. It

---

* The visual cortex is in the occipital lobe, at the back of the brain.

could equally be that a mature science of consciousness will allow us to explain, predict, and control phenomenological properties without ever delivering the intuitive feeling that "yes, this is right, *of course* it has to be this way!"

Importantly, the real problem of consciousness is not an admission of defeat to the hard problem. The real problem goes after the hard problem indirectly, but it still goes after it. To understand why this is so, let me introduce the "neural correlates of consciousness."

IT STILL AMAZES ME how disreputable consciousness science was, even just thirty years ago. In 1989, one year before I started my undergraduate degree at Cambridge University, the leading psychologist Stuart Sutherland wrote: "Consciousness is a fascinating but elusive phenomenon. It is impossible to specify what it is, what it does, or why it evolved. Nothing worth reading has been written on it." This damning summary appeared in no lesser place than the *International Dictionary of Psychology*, and it captures the attitude toward consciousness that I often encountered in my first steps into academia.

Elsewhere, far away from Cambridge and though I did not know it at the time, the situation was more promising. Francis Crick (the co-discoverer, with Rosalind Franklin and James Watson, of the molecular structure of DNA) and his colleague Christof Koch, who were both based in San Diego in California, were setting out what would become the dominant method in the rise of consciousness science—the search for the neural correlates of consciousness.

The gold-standard definition of a neural correlate of consciousness, or NCC, is "the minimal neuronal mechanisms jointly sufficient for any one specific conscious percept." The NCC approach proposes that there is some specific pattern of neural activity that is responsible for any and every experience, such as the experience of "seeing red." Whenever this activity is present, an experience of redness will happen, and whenever it isn't, it won't.

The great merit of the NCC approach is that it offers a practical recipe for doing research. To identify an NCC, all you need to do is concoct a situation in which people sometimes have a particular conscious experience, and at other times do not, while making sure that these conditions are otherwise as closely matched as possible. Given such a situation, you then compare activity in the brain between the two conditions, using brain imaging methods such as functional magnetic resonance imaging (fMRI) or electroencephalography (EEG).* The brain activity specific to the "conscious" condition reflects the NCC for that particular experience.

The phenomenon of "binocular rivalry" offers a helpful example. In binocular rivalry, a different image is shown to each eye—perhaps a picture of a face to the left eye and a picture of a house to the right eye. In this situation, conscious perception doesn't settle on a weird face-house chimera. It flips back and forth between the face and the house, dwelling for a few seconds on each. First you see a house, then a face, then a house again . . . and so on. What's important here is that conscious perception changes even though the sensory input remains constant. By looking at what happens in the brain, it's therefore possible to distinguish brain activity that tracks conscious perception from activity that tracks whatever the

---

* Functional MRI (fMRI) measures a metabolic signal (blood oxygenation) related to neural activity—it offers high spatial detail but is only indirectly related to what neurons do. EEG measures the tiny electrical signals generated by the activity of large populations of neurons near the cortical surface. This method tracks brain activity more directly than fMRI, but with lower spatial specificity.

sensory input happens to be. The brain activity that goes along with the conscious perception identifies the NCC for that perception.

The NCC strategy has been impressively productive over many years, delivering reams of fascinating findings, but its limitations are becoming apparent. One problem is that it is difficult, and perhaps in the end impossible, to disentangle a "true" NCC from a range of potentially confounding factors, the most important of which are those neural happenings that are either prerequisites for, or consequences of, an NCC itself. In the case of binocular rivalry, brain activity that goes along with the conscious perception may also track upstream (prerequisite) processes like "paying attention" and, on the downstream side, the verbal behavior of "reporting"—of saying that you see a house or a face. Although related to the flow of conscious perception, the neural mechanisms responsible for attention and verbal report—or other prerequisites and downstream consequences—should not be confused with those that are responsible for the conscious perception itself.

The deeper problem is that *correlations* are not *explanations*. We all know that mere correlation does not establish causation, but it is also true that correlation falls short of explanation. Even with increasingly ingenious experimental designs and ever more powerful brain imaging technologies, correlation by itself can never amount to explanation. From this perspective, the NCC strategy and the hard problem are natural bedfellows. If we restrict ourselves to collecting correlations between things happening in the brain and things happening in our experience, it is no surprise that we will always suspect an explanatory gap between the physical and the phenomenal. But if we instead move beyond establishing correlations to discover explanations that connect properties of neural mechanisms to properties of subjective experience, as the real problem approach advocates, then this gap will narrow and might even disappear entirely. When we are able to predict (and explain, and control) why the experience of redness is the particular way it is—and not like blueness, or like jealousy—the mystery of how redness happens will be less mysterious, or perhaps no longer mysterious at all.

The ambition of the real problem approach is that as we build ever sturdier explanatory bridges from the physical to the phenomenological, the hard problem intuition that consciousness can never be understood in physical terms will fade away, eventually vanishing in a puff of metaphysical smoke. When it does, we will have in our hands a satisfactory and fully satisfying science of conscious experience.

What justifies this ambition? Consider how the scientific understanding of *life* has matured over the last century or two.

NOT SO LONG AGO, life seemed as mysterious as consciousness does today. Scientists and philosophers of the day doubted that physical or chemical mechanisms could ever explain the property of *being alive*. The difference between the living and the nonliving, between the animate and the inanimate, appeared so fundamental that it was considered implausible that it could ever be bridged by mechanistic explanations of any sort.

This philosophy of *vitalism* reached a peak in the nineteenth century. It was supported by leading biologists like Johannes Müller and Louis Pasteur, and it persisted well into the twentieth century. Vitalists thought that the property of being alive could be explained only by appealing to some special sauce: a spark of life, an *élan vital*. But as we now know, no special sauce is needed. Vitalism today is thoroughly rejected in scientific circles. Although there are still many things about life that remain unknown—how a cell works, for example—the idea that being alive requires some supernatural ingredient has lost all credibility. The fatal flaw of vitalism was to interpret a failure of imagination as an insight into necessity. This is the same flaw that lies at the heart of the zombie argument.

The science of life was able to move beyond the myopia of vitalism, thanks to a focus on practical progress—to an emphasis on the "real problems" of what being alive means. Undeterred by vitalistic pessimism, biologists got on with the job of describing the properties of living systems, and then explaining (also predicting and controlling) each of these properties in terms of physical and chemical mechanisms. Reproduction, metabolism, growth, self-repair, development, homeostatic self-regulation—all became individually and collectively amenable to mechanistic explanation. As the details became filled in—and they are still being filled in—not only did the basic mystery of "what is life" fade away, the very concept of life ramified so that "being alive" is no longer thought of as a single all-or-nothing property. Gray areas emerged, famously with viruses but now also with synthetic organisms and even collections of oil droplets, each of which possess some but not all of the characteristic properties of living systems. Life became naturalized and all the more fascinating for having become so.

This parallel provides both a source of optimism and a practical strategy for addressing the real problem of consciousness.

The optimism is that today's consciousness researchers may be in a situation similar to that facing biologists, studying the nature of life, just a few generations ago. What counts as mysterious *now* may not *always* count as mysterious. As we get on with explaining the various properties of consciousness in terms of their underlying mechanisms, perhaps the fundamental mystery of "how consciousness happens" will fade away, just as the mystery of "what is life" also faded away.

Of course, the parallel between life and consciousness is not perfect. Most conspicuously, the properties of life are objectively describable, whereas the explanatory targets of consciousness science are subjective—they exist only in the first person. However, this is not an insurmountable barrier; it mostly means the relevant data, because they are subjective, are harder to collect.

The practical strategy stems from the insight that consciousness, like life, is not just one single phenomenon. By shifting the focus away from life as one big scary mystery, biologists became less inclined to desire, or to require, one humdinger eureka of a solution. Instead, they divided the "problem" of life into a number of related but distinguishable processes. Applying the same strategy to consciousness, in this book I will focus on *level*, *content*, and *self* as the core properties of what *being you* is all about. By doing so, a fulfilling picture of all conscious experience will come to light.

Conscious *LEVEL* CONCERNS "how conscious we are"—on a scale from complete absence of any conscious experience at all, as in coma or brain death, all the way to vivid states of awareness that accompany normal waking life.

Conscious *content* is about what we are conscious of—the sights, sounds, smells, emotions, moods, thoughts, and beliefs that make up our inner universe. Conscious contents are all varieties of *perception*—brain-based interpretations of sensory signals that collectively make up our conscious experiences. (Perception, as we will see, can be both conscious and unconscious.)

Then there's conscious *self*—the specific experience of *being you*, and the guiding theme of this book. The experience of "being a self" is a subset of conscious contents, encompassing experiences of having a particular body, a first-person perspective, a set of unique memories, as well as experiences of moods, emotions, and "free will." Selfhood is probably the aspect of consciousness that we cling to most tightly, so tightly that it can be tempting to confuse self-consciousness (the experience of being a self) with consciousness itself (the presence of any kind of subjective experience, of any phenomenology, whatsoever).

In making these distinctions, I am not proposing that these aspects of consciousness are completely independent. In fact, they are not, and figuring out how they relate presents another significant challenge for consciousness science.

Nonetheless, dividing up the real problem of consciousness in these broad terms has many benefits. By providing distinct targets for explanation, it becomes more feasible to propose possible mechanisms able to do the required jobs of explanation, prediction, and control. Equally important, it pushes back against the limiting idea that consciousness is just "one thing"—a single intimidating mystery that might elude scientific explanation altogether. We will instead see how different properties of consciousness come together in different ways, across species and even among different people. There are as many different ways of being conscious as there are different conscious organisms.

Eventually, the hard problem itself may succumb, so that we will be able to understand consciousness as being continuous with the rest of nature without having to adopt any arbitrary "ism" stating by fiat how phenomenology and physics are related.

This is the promise of the real problem. To see how far it can take us, read on.

2

MEASURING CONSCIOUSNESS

How conscious are you, right now? What makes the difference between being conscious at all and being a chunk of living meat, or lifeless silicon, without any inner universe? New theories and technologies are allowing scientists, for the first time, to measure *levels* of consciousness. To best understand this new research, let's look at the roots of its development.

In seventeenth-century Paris, there was a cellar, deep, dark, and cool beneath the Observatoire, on the left bank of the Seine. This cellar played a surprising role in the history of science—a role that showcases the importance of *measurement* in the advance of knowledge.

Philosophers and scientists of the day, though they were not yet called scientists, were racing to develop reliable thermometers, and by doing so to reach a physical understanding of the nature of heat. Popular "calorific" theories, according to which heat was a substance that could flow into and out of objects, were falling out of favor. Revising these theories depended on precise experiments in which the "hotness" or "coolness" of objects could be systematically assessed. Such experiments needed both a means of measuring whatever "heat" was and a scale on which different measurements could be compared. The race was on to develop a reliable *thermometer* and a scale of *temperature*. But how could the reliability of a thermometer be established, if not against a well-validated scale? And how to develop a temperature scale without already having a reliable thermometer?

The first step in solving this conundrum was to come up with a fixed point: an unchanging reference that could be assumed to have constant temperature. Even this was challenging. Promising candidates such as the boiling point of water were known to depend on factors like air pressure, which varied with altitude and weather, and even on subtle influences like the surface roughness of a glass vessel. It was because of frustrations like this that, for a while, a Parisian cellar—with its apparently constant coolness—seemed a reasonable choice for the fixed point of temperature. (This was not the only unusual suggestion. Perhaps the strangest came from a certain Joachim Dalencé, who suggested the melting point of butter.)

Eventually, reliable and precise mercury-based thermometers were invented, which led to calorific theories being superseded by a new science of thermodynamics—a revolution associated with legendary figures such as Ludwig Boltzmann and Lord Kelvin. In thermodynamics, temperature is a large-scale property of the movement of the molecules within a substance; specifically, the mean molecular kinetic energy. Faster movement, higher temperature. "Heat" becomes the energy transferred between two systems at different temperatures. Importantly, thermodynamics did more than merely establish that mean kinetic energy *correlated* with temperature—it proposed that this is what temperature actually *is*. Armed with this new theory, scientists could now talk about the temperature at the surface of the sun, and even identify an "absolute zero" at which all molecular movement, in theory, ceased. Early scales based on measurements of particular substances (Celsius, Fahrenheit) were replaced by a scale based on underlying physical properties (Kelvin, named after the lord). The physical basis of temperature and heat is no longer a mystery.

I first read this story in the book *Inventing Temperature* by the historian Hasok Chang of University College London. Until then, I'd never fully appreciated the extent to which scientific progress depends on measurement. The history of thermometry, and its impact on our understanding of heat, offers a vivid example of how the ability to make detailed quantitative measurements, on a scale defined by fixed points, has the power to transform something mysterious into something comprehensible.

Could the same approach work for consciousness?

PHILOSOPHERS SOMETIMES TALK ABOUT a hypothetical "consciousness meter" which is able to determine whether something—another person, animal, or perhaps machine—is conscious or not. At a conference in the 1990s, in the heyday of the hard problem, David Chalmers took an old hair dryer and pointed it at his head to emphasize how useful such a thing would be, were it to exist. Point your consciousness meter at something and read off the answer. No more mystery about how far the charmed circle of consciousness extends.

As the temperature story shows, though, the value of measurement lies not only in delivering yes/no answers about the presence or absence of a property, but in making possible the quantitative experiments that have the potential to transform scientific understanding.

If consciousness turns out to be something like temperature—which is to say, if there is a single physical process that underlies and is identical to "being conscious"—the payoff would be spectacular. Not only would we be able to determine "how conscious" someone is, we would be able to talk sensibly about specific "levels" and "degrees" of consciousness, and about varieties of consciousness far removed from our parochial human example.

But even if the story of consciousness turns out differently, to be less like temperature and more like life, as I suspect it will, the ability to make precise measurements nevertheless remains an essential step in building explanatory bridges; in explaining, predicting, and controlling the nature of subjective experiences. In either scenario, measurement turns the qualitative into the quantitative, the vague into the precise.

Measurement has practical motivations too. The art of anesthesia—applied to more than four million people every day—involves maintaining a patient in a temporary oblivion without overdosing them. A reliable and precise consciousness meter would be of obvious value in achieving this delicate balance, especially since anesthesia is frequently accompanied by neuromuscular blocking agents which induce a temporary paralysis, allowing surgeons to do their work unencumbered by muscular reflexes. And, as we'll soon see, there is an urgent need for new methods of deciding whether consciousness remains after severe brain damage, when patients are given terrifying diagnoses like the "vegetative state" or the "minimally conscious state."

In fact, brain-based consciousness monitors have been deployed in operating theaters for many years already. The most common is the "bispectral index" monitor. While the details lie hidden beneath patents, the basic concept is to combine a range of electroencephalographic (EEG) measures together into a single continuously updated number, to guide the anesthesiologist during surgery. This is a fine idea, but bispectral index monitors have remained controversial, partly because there have been several instances where their readings have been inconsistent with other behavioral signs of consciousness, such as patients opening their eyes or remembering what the surgeons had been saying during an operation. A deeper problem, when it comes to consciousness science, is that the bispectral index is not based on any principled theory.

Over the last few years, a new generation of consciousness meters has started to take shape—not in operating theaters but in neuroscience laboratories. Unlike previous consciousness monitors, these new approaches are tied closely to an emerging theoretical understanding of the brain basis of being conscious, and they are already showing their practical worth.

MEASURING CONSCIOUS LEVEL in humans is not the same as deciding whether someone is awake or asleep. Conscious level is not the same thing as physiological arousal. While the two are often highly correlated, consciousness (awareness) and wakefulness (arousal) can come apart in various ways, which is enough to show that they cannot depend on the same underlying biology. When you are dreaming you are by definition asleep, but you are having rich and varied conscious experiences. At the other extreme lie catastrophic conditions like the vegetative state (also now known as "unresponsive wakefulness syndrome"), in which a person still cycles through sleep and wakefulness, but shows no behavioral signs of conscious awareness: the lights are occasionally on, but there's nobody home. The figure on the next page illustrates the relation between awareness and wakefulness across a variety of different conditions—both normal and pathological.

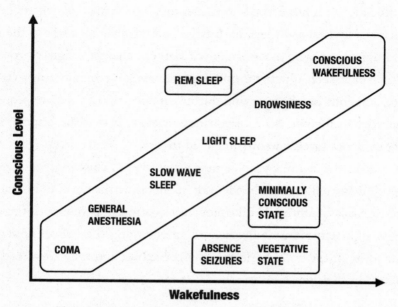

The relationship between conscious level (awareness) and wakefulness (arousal).

To track conscious level, we need to ask what in the brain underlies being conscious, as opposed to merely being awake. Could it simply be the number of neurons involved? It doesn't seem so. The cerebellum (the "little brain" hanging off the back of your cortex) has about four times as many neurons as the rest of the brain put together, but seems barely involved in consciousness. There is a rare condition called cerebellar agenesis in which people fail to develop anything like a normal cerebellum yet still manage to lead largely normal lives. Certainly, there is no reason to doubt that they are conscious.

What about the overall degree of neuronal activity? Is the brain generally more active in conscious states than in unconscious states? Well, maybe, to some extent—but not by much. Although there are differences in the brain's energy consumption across conscious levels, these differences are rather small, and certainly there is no sense in which the brain "shuts down" when consciousness fades.

Consciousness instead seems to depend on how different parts of the brain speak to each other. And not the brain as a whole: the activity patterns that matter seem to be those within the thalamo-cortical system—the combination of the cerebral cortex and the thalamus (a set of oval-shaped brain structures—"nuclei"—sitting just below, and intricately connected with, the cortex). The latest and most exciting approaches to measuring conscious level—and distinguishing it from wakefulness—are based on tracking and quantifying these interactions. The most ambitious version of this idea delivers a single number that indicates how conscious a person is. Just like a thermometer.

THIS NEW APPROACH HAS BEEN pioneered by the Italian neuroscientist Marcello Massimini, initially with the renowned consciousness researcher Giulio Tononi at the University of Wisconsin–Madison, and more recently with his own group at the University of Milan. What they did was simple and elegant. To test how different parts of the cortex were talking to each other, they stimulated activity in one location and recorded how this pulse of activity spread to other cortical regions over space and time. They did this by combining two techniques: EEG and transcranial magnetic stimulation (TMS). A TMS rig is a precisely controlled electromagnet which allows a researcher to inject a short and sharp pulse of energy directly into the brain through the skull, while EEG in this case is used to record the brain's response to this zapping. It's like banging on the brain with an electrical hammer and listening to the echo.

Perhaps surprisingly, people are rarely aware of the TMS zap itself, unless it does something obvious like eliciting a movement (when the magnet is placed over the motor cortex, which controls actions) or a simple visual flash (a "phosphene"—which can happen when the visual cortex is activated). And if the zap causes the muscles in your face and scalp to spasm, you'll notice the pain. But for the most part, the huge disturbance in brain activity caused by TMS does not generate any alteration in conscious experience at all. Maybe this is not so surprising. It just shows we are not aware of what our neurons are doing—and why should we be?

Even though we do not directly feel the TMS pulses, Massimini and Tononi found that their electrical echoes could be used to distinguish different levels of consciousness. In unconscious states, like dreamless sleep and general anesthesia, these echoes are very simple. There is a strong initial response in the part of the brain that was zapped, but this response dies away quickly, like the ripples caused by throwing a stone into still water. But during conscious states, the response is very different: a typical echo ranges widely over the cortical surface, disappearing and reappearing in complex patterns. The complexity of these patterns, across space and time, implies that different parts of the brain—in particular the thalamocortical system—are communicating with each other in much more sophisticated ways during conscious states than during unconscious states.

While the difference between the two conditions is often easy to see simply by eyeballing the data, what's truly exciting about this work is that the complexity of the echo can be quantified. It is possible to put a number to it to specify the *magnitude* of complexity. The approach is called "zap and zip": use TMS to zap the cortex, and use a computer algorithm to "zip" the response, the electrical echo, into a single number.

The algorithm used by the "zip" part is the same one used to compress (zip) digital photos into smaller files. Any pattern, whether a photo of your summer holiday or an electrical echo unfolding across the brain in time and space, can be represented as a sequence of 1s and 0s. For any nonrandom sequence there will be a compressed representation, a much shorter string of numbers that can be used to fully regenerate the original. The length of the shortest possible compressed representation is called the "algorithmic complexity" of the sequence. Algorithmic complexity will be lowest for a completely predictable sequence (such as a sequence consisting entirely of 1s, or of 0s), highest for a completely random sequence, and somewhere in the middle for sequences that contain some amount of predictable structure. The "zip" algorithm—which calculates what's called "Lempel-Ziv-Welch complexity," or "LZW complexity" for short—is a popular way of estimating the algorithmic complexity for any given sequence.

Massimini and his team called their measure of the echoes recorded in their experiments the "perturbational complexity index", or PCI. It uses LZW complexity to provide a measure (index) of the algorithmic complexity of the brain's response to a perturbation—the TMS pulse.

They first validated their measure by showing that PCI values during unconscious states, such as dreamless sleep and general anesthesia, were reliably lower than during a baseline conscious state of resting wakefulness. This is reassuring, but the real power of the PCI approach is that it defines a continuous scale, allowing more fine-grained distinctions to be made. In a landmark study from 2013, Massimini's team measured the PCI values of a large number of patients with brain injuries who had disorders of consciousness. They found that PCI magnitudes correlated extremely well with levels of impairment, as independently diagnosed by neu-rologists. For example, people in a vegetative state, where conscious-ness is assumed to be absent despite preserved wakefulness, had lower PCI scores than those in minimally conscious states, in which behavioral signs of consciousness come and go. They were even able to draw a dividing line between PCI values indicative of conscious-ness from those suggesting its absence.

In my research group at the University of Sussex, we've been working on similar methods to assess conscious level. But instead of using TMS to inject pulses of energy into the cortex, we've been measuring the algorithmic complexity of ongoing, natural—what we call "spontaneous"—brain activity. Think of it as "zipping" with-out the "zapping." In a series of studies, led by my colleague Adam Barrett and our former PhD student Michael Schartner, we discov-ered that the complexity of spontaneous cortical activity—as mea-sured by EEG—reliably drops in both early sleep and anesthesia. We also found that complexity during rapid eye movement (REM) sleep is much the same as during normal conscious wakefulness, which makes sense because REM sleep is when dreaming is most likely—and dreams are conscious. Massimini and his team had found the same pattern of results with their PCI measure, further supporting the claim that these measures are tracking conscious level rather than wakefulness.

THE ABILITY TO MEASURE conscious level independently from wakefulness is not just scientifically important, it is potentially game-changing for neurologists, and for their patients. Massimini's 2013 study already demonstrates that the PCI can distinguish between the vegetative state and the minimally conscious state.

Measures like the PCI are so powerful in this context because they do not rely on outwardly visible behavior. Mere wakefulness—physiological arousal—is defined in terms of behavior. In the clinic, neurologists typically infer wakefulness when a person responds to sensory stimulation, like a loud noise or a pinch on the arm. But consciousness is defined in terms of inner subjective experience, and so can only ever be indirectly related to what can be seen from the outside.

Standard clinical approaches to determining the conscious status of a brain-injured patient still rely on behavior. Typically, neurologists assess whether a patient not only responds to sensory stimulation—the marker of physiological arousal—but is able to interact with their environment, by responding to commands, or by engaging in voluntary behavior. When a patient can obey a two-part request and clearly state their name and the date, we infer full consciousness. The problem with this approach is that some patients may still possess an inner life but be unable to express it outwardly. Inferences based purely on behavior will miss these cases, diagnosing absence of consciousness when in fact consciousness remains.

An extreme example is "locked-in syndrome," where consciousness is fully present despite total paralysis of the body. This rare affliction can follow damage to the brainstem, a region at the base of the brain (and at the top of the spinal cord) which, among other roles, mediates control of muscles in the body and in the face. Thanks to a quirk of anatomy, locked-in patients may still retain the ability to make limited eye movements, opening a narrow and easily missed behavioral channel for diagnosis and communication. A former editor of *Elle* magazine—Jean-Dominique Bauby, who became locked-in in 1995 following a brain hemorrhage—wrote an entire book this way, *The Diving Bell and the Butterfly*. So-called

"complete" locked-in patients lack even these communication chan-
nels, making diagnosis even harder. When relying on behavior
alone, it is all too easy to mistake locked-in syndrome for the complete and permanent absence of consciousness. But put someone
like Bauby inside a brain scanner and it will be easy to see that their
overall brain activity is almost completely normal. In Massimini's
2013 study, locked-in patients had PCI values indistinguishable
from healthy age-matched controls—indicating fully intact consciousness.

The more challenging cases arise in the gray zone between life
and death, in conditions like the vegetative state and the minimally
conscious state. In these borderlands, behavioral signs of consciousness can be absent or inconsistent, and damage to the brain so widespread that brain scans may also be inconclusive. It is here that
measures like PCI could be truly game-changing. When a patient
has a PCI score that suggests consciousness, even if everything
else about them suggests otherwise, then this is a patient worth a
second look.

Marcello Massimini recently told me about one case where
measuring PCI made all the difference. A young man with severe
head injuries had been admitted to a hospital in Milan. He remained
unresponsive to simple questions and commands, suggesting a diagnosis of vegetative state. But his PCI was as high as a healthy,
fully conscious person's—a perplexing observation, especially since
he was not locked-in. The clinical team eventually tracked down a
relative, an uncle who traveled to Italy from North Africa, where
the young man's family still lived. When this uncle started engaging with his nephew in Arabic there was an instant response: smil-
ing at jokes, even giving a thumbs-up when watching a movie.
He had been conscious all along—just unresponsive *in Italian*.

Why this should be is hard to say. Massimini believes it might be a strange case of "cultural neglect," as if the Italian world simply did not matter to him anymore. Either way, this young man's story could have ended very differently without the telltale electrical echoes measured by the PCI.

The diagnosis of residual consciousness in brain-injured patients is a fast-moving area of medicine. Along with Massimini's PCI, several other methods are now migrating from the lab to the clinic. My favorite is based on the famous—in neurology circles—"house tennis" experiment conducted by the neuroscientist Adrian Owen and his team in 2006. In Owen's experiment, a twenty-three-year-old woman, behaviorally unresponsive following a traffic accident, was placed in an fMRI scanner and given a series of verbal instructions. Sometimes she was asked to imagine playing tennis, while at other times she was invited to imagine walking around the rooms of her house. On the face of it, this seems a peculiar thing to do, since patients like this are not responsive to anything—let alone to complex verbal instructions. However, studies with healthy people have shown that the brain regions engaged by imagining fluent movements (like playing tennis) are highly distinct from those activated by imagining navigating through spaces.* Remarkably, Owen's patient showed exactly the same pattern of brain responses, indicating that she, too, was actively following the instructions by engaging in highly specific mental imagery. It is almost impossible to conceive that anyone could do this while unconscious, so Owen concluded

---

* Imagining (and executing) fluent movements activates cortical regions such as the supplementary motor area, while imagining spatial navigation lights up other regions, such as the parahippocampal gyrus. Anatomically, these brain areas are quite distant from each other. Unsurprisingly, both imagery tasks activate regions involved in hearing and in language processing.

that the behavioral diagnosis of vegetative state was wrong, and that the young woman was in fact conscious. In effect, Owen and his team had repurposed a brain scanner to allow his patient to interact with her environment using her brain rather than her body.

Subsequent studies have gone further still, using Owen's method not only for diagnosis but also for communication. In a 2010 study led by Martin Monti, a patient who had been admitted with a diagnosis of vegetative state was able to answer yes/no questions by imagining playing tennis for "yes" and imagining walking around their house for "no." A laborious way of communicating, for sure, but a life-changing development for those with no other way to make themselves understood.

How many unresponsive but conscious people might languish forgotten in neurology wards and nursing homes? It is difficult to know. Owen's method—being older than PCI—has been investigated more often, with a recent analysis suggesting that between 10 and 20 percent of vegetative state patients might retain some form of covert consciousness, a number which would translate into many thousands across the world. And this may well be an underestimate. To pass the Owen test, patients still have to understand language and engage in extended periods of mental imagery, which some may not be able to do, despite being conscious. It is here that new methods like PCI are especially significant, since they promise the ability to detect residual awareness without requiring the patient to do anything at all. Just as a true consciousness meter should.

THE CONCEPT OF "level of consciousness," as I've used it so far, picks out relatively global changes in how conscious an individual is—such as the difference between normal waking life and general anesthesia, or being in a vegetative state. However, there are other ways to think about what conscious level might mean. Is a baby less conscious than an adult? Is a tortoise less conscious than either?

There is of course a danger in thinking along these lines. Such questions seduce us into presuming that any form of consciousness that diverges from the healthy adult human is somehow lesser, or lower. This way of thinking is symptomatic of the human exceptionalism that has repeatedly plagued biology, as it has darkened the history of human thought everywhere. Consciousness has many properties, and it is a mistake to confuse the expression of the particular set of properties typical of healthy adult humans with the essential nature of consciousness in all its forms, and then to assume that it sits at the top of a unidimensional scale. Conscious experiences surely emerge over time, whether in the development of any single animal (human or not) or across the vast expanse of evolution. But it's a considerable leap to describe either process as unfolding along a single line, or as culminating in the adult human ideal of what it's like to be you, or to be me. This is one way in which the analogy between consciousness and temperature—with which I began this chapter—may be limited.

A related question is whether consciousness is "all or none"—either the lights are on or they aren't—or whether it is "graded," with no bright line between consciousness and its absence. This question applies equally to the emergence of consciousness in evolution or in development, as to when returning from the oblivion of anesthesia or dreamless sleep. Although this question is alluring, I think it is misguided. The distinction between "all or none" and "graded" consciousness doesn't have to be either-or. Whether in evolution, in development, in daily life, or in the neurology ward, I prefer to think in terms of sharpish transitions from the total absence of consciousness to the presence of at least some conscious experience, with conscious experience then manifesting in different degrees, perhaps along different dimensions, once the inner lights are at least glimmering.

Take a typical adult human. Is her level of consciousness higher (or lower) when she is dreaming than when she is sitting at her desk after a heavy lunch, in a distracted semi-stupor? There are no straightforward answers to questions like these. Dreaming may be "more conscious" in some ways (for example, the vividness of perceptual phenomenology) but "less conscious" in others (for example, the degree of reflective insight into what is happening).*

An important consequence of taking multidimensional levels of consciousness seriously is that sharp distinctions between conscious *level* and conscious *content* disappear. It becomes meaningless to completely separate how conscious you are from what you are conscious of. A "one size fits all" measure of consciousness—of the sort we might expect if we take the temperature analogy too literally—may never be enough.

One example of how conscious level and conscious content interact comes from a study we carried out a few years ago on brain activity in the psychedelic state. Among their many uses, psychedelic drugs offer unique opportunities for consciousness science since they induce profound alterations in conscious contents, arising from a simple pharmacological intervention in the brain.

---

* Reflective insight is preserved in the rare "lucid dreaming" state, in which dreamers are aware that they are dreaming and can voluntarily direct their behavior. In a remarkable recent study, researchers were able to communicate with people during lucid dreams by using their eye movements as a channel, much like the locked-in patients described earlier. These dreamers were able to correctly answer simple math problems, as well as various yes/no questions.

The Swiss chemist Albert Hofmann, the inventor of lysergic acid diethylamide—LSD—gives a sense of how dramatic these alterations can be in his record of his journey home from the laboratories of the pharmaceutical company Sandoz, in Basel, on 19 April 1943. On this day, which is now remembered as "Bicycle Day," he had decided to swallow a small amount of his recent discovery. Shortly afterward, he began to feel rather unusual, and so set off home on his bike. After somehow making it back while being assailed by all manner of distressing experiences, and having believed he was going insane, he lay down on his sofa and closed his eyes.

> . . . Little by little I could begin to enjoy the unprecedented colors and plays of shapes that persisted behind my closed eyes. Kaleidoscopic, fantastic images surged in on me, alternating, variegated, opening and then closing themselves in circles and spirals, exploding in colored fountains, rearranging and hybridizing themselves in constant flux . . .

In the psychedelic state, vivid perceptual hallucinations are frequently accompanied by unusual experiences of selfhood often described as "ego dissolution," where the boundaries between self and world, and other people, appear to shift or dissolve. These departures from "normal" conscious experience are so pervasive that the psychedelic state might represent not only a change in conscious *contents*, but also a change in overall conscious *level*. This is the idea we set out to test, in a collaboration with Robin Carhart-Harris at Imperial College London and Suresh Muthukumaraswamy at the University of Auckland.

In April 2016, Robin and I were at a conference in the foothills of the Santa Catalina Mountains, just outside Tucson, in Arizona. We'd both been invited to give talks about our research, and we were using the opportunity to explore how our interests in consciousness might overlap in the context of psychedelics. Scientific and medical research on LSD, and on other psychedelic compounds like psilocybin (the active ingredient in magic mushrooms), had only recently restarted after decades in the wilderness. Following Hofmann's self-experimentation, there had been a brief flowering of studies exploring the potential of LSD for treating a range of psychological disorders, including addiction and alcoholism, with very promising results. But the subsequent uptake of LSD as a recreational drug and as a symbol of rebellion, evangelized by Timothy Leary among others, led to pretty much all of this research being shut down by the end of the 1960s. It took until the 2000s before any substantial new research restarted—a lost generation of scientific progress.

At the level of neurochemistry, the classic psychedelics—LSD, psilocybin, mescaline, and dimethyltryptamine (DMT, the active ingredient in the South American hallucinogenic brew called ayahuasca)—work primarily by affecting the brain's serotonin system. Serotonin is one of the brain's primary neurotransmitters—chemicals which wash through the brain's circuits and which influence how neurons communicate. Psychedelic drugs influence the serotonin system by binding strongly to a specific serotonin receptor, the 5-HT_{2A} receptor, which is found throughout much of the brain. One of the main challenges for psychedelic research is to understand how these low-level pharmacological interventions alter global patterns of brain activity so as to deliver profound changes in conscious experience.

Robin's team had previously discovered that the psychedelic state involves striking alterations in brain dynamics, when compared to a placebo control condition. Networks of brain regions that are usually coordinated in their activity—so-called "resting-state networks"—become uncoupled, and other regions that are usually more or less independent become linked. Overall, the picture is of a breakdown in the patterns of connectivity that characterize the brain under normal conditions. Robin's idea was that these breakdowns could account for signature features of the psychedelic state, like the dissolution of boundaries between self and world, and the intermingling of the senses.

Robin and I realized that the data his team been collecting were ideally suited for the algorithmic complexity analyses we'd been applying, with my team at Sussex, to sleep and anesthesia. In particular, some of their brain scans had been carried out using magneto-encephalography (MEG), which provides the high time resolution and global brain coverage that we needed. They had used MEG to measure brain activity in volunteers who had taken psilocybin, LSD, or low doses of ketamine. (While high doses of ketamine act as an anesthetic, low doses have more of a hallucinogenic effect.) We could use these data to answer the question: What happens to measures of conscious level when conscious contents change as dramatically as they do on a psychedelic trip?

Back at Sussex, Michael Schartner and Adam Barrett calculated the changes in the algorithmic complexity of the MEG signal across many different regions in the brain for all three psychedelic states. The results were clear and surprising: psilocybin, LSD, and ketamine all led to *increases* when compared to a placebo control. This was the first time anyone had seen an increase in a measure of conscious level relative to a baseline of waking rest. All previous comparisons whether through sleep or anesthesia or disorders of consciousness, had led to decreases in these measures.

To understand what this result means, remember that the

measures of algorithmic complexity we used are best thought of as measures of the randomness, or "signal diversity," of the brain signals that they are applied to. A fully random sequence will have the highest possible algorithmic complexity, the greatest possible diversity. Our findings therefore complemented Robin's previous studies by showing that brain activity in the psychedelic state becomes more random over time, in line with the freewheeling reorganization of perceptual experience that people frequently report during a trip. They also shed new light on how conscious level and conscious content relate. Here is an example of a measure of conscious *level* responding to the widespread changes in the *contents* of consciousness that characterize the psychedelic state. The fact that a measure of conscious level is also sensitive to changes in conscious content makes clear that they are not independent aspects of consciousness.

The results from our psychedelic analyses raised a disturbing prospect. Would maximally random brain activity, as measured by algorithmic complexity, lead to a maximally psychedelic experience? Or to a different "level" of consciousness of some other kind? The extrapolation seems unlikely. A brain with all its neurons firing willy-nilly would seem more likely to give rise to no conscious experience at all, just as free-form jazz at some point stops being music.

The issue here is that algorithmic complexity is a poor approximation of what "being complex" usually means. Intuitively, complexity is not the same as randomness. A more satisfying notion of complexity is as the middle ground between order and disorder— not the extreme point of disorder. It is Nina Simone and Thelonious Monk, not the Bonzo Dog Doo-Dah Band.* What happens if we

---

* On their 1967 debut album *Gorilla*, the Bonzo Dog Doo-Dah Band parodied traditional jazz music by attempting to play as badly as possible. There'll be more about gorillas later in this book.

run with this more sophisticated way of thinking about complexity instead?

A PAPER PUBLISHED IN 1998 by Giulio Tononi and my former boss and mentor Gerald Edelman, in the journal *Science*, does exactly this. I still remember reading this paper, now more than twenty years ago. It was a landmark event in my thinking about consciousness, and a large part of what drew me to work at the Neurosciences Institute in San Diego.

Instead of focusing, in the style of the "neural correlates of consciousness" (NCC) approach, on a single exemplary conscious experience—like the experience of "seeing the color red"—Tononi and Edelman asked what was characteristic about conscious experiences in general. They made a simple but profound observation: that conscious experiences—*all* conscious experiences—are both *informative* and *integrated*. From this starting point, they made claims about the neural basis of *every* conscious experience, not just of specific experiences of seeing red, or feeling jealous, or suffering a toothache.

The idea of consciousness as simultaneously informative and integrated needs a little unpacking.

Let's start with information. What does it mean to say that conscious experiences are "informative"? Edelman and Tononi did not mean this in the sense that reading a newspaper can be informative, but in a sense that, though it might at first seem trivial, conceals a great deal of richness. Conscious experiences are informative because every conscious experience is different from every other conscious experience that you have ever had, ever will have, or ever could have.

Looking past the desk in front of me through the window beyond, I have never before experienced precisely *this* configuration of coffee cups, computer monitors, and clouds—an experience that is even more distinctive when combined with all the other perceptions, emotions, and thoughts that are simultaneously present in the background of my inner universe. At any one time, we have precisely *one* conscious experience out of *vastly many possible* conscious experiences. Every conscious experience therefore delivers a massive reduction of uncertainty, since *this* experience is being had, and not *that* experience, or *that* experience, and so on. And reduction of uncertainty is—mathematically—what is meant by "information."

The informativeness of a particular conscious experience is not a function of how rich or detailed it is, or of how enlightening it is to the person having that experience. Listening to Nina Simone while eating strawberries on a roller coaster rules out just as many alternative experiences as does sitting with eyes closed in a silent room, experiencing close to nothing. Each experience reduces uncertainty with respect to the range of possible experiences by the same amount.

In this view, the "what-it-is-like-ness" of any specific conscious experience is defined not so much by what it *is*, but by all the unrealized but possible things that it is *not*. An experience of pure redness is the way that it is, not because of any intrinsic property of "redness," but because red is not blue, green, or any other color, or any smell, or a thought or a feeling of regret or indeed any other form of mental content whatsoever. Redness is redness because of all the things it isn't, and the same goes for all other conscious experiences.

Scoring high on information is not by itself enough. Conscious experiences are not only highly informative, they are also *integrated*. Exactly what is meant by consciousness being "integrated" is still much debated, but essentially it means that every conscious experience appears as a unified scene. We do not experience colors separately from their shapes, nor objects independently of their background. The many different elements of my conscious experience right now—computers and coffee cups, as well as the sound of a door closing in the hallway and my thoughts about what to write next—seem tied together in an inescapable and fundamental way, as aspects of a single encompassing conscious scene.

The key move made by Tononi and Edelman was to propose that if every conscious experience is both informative and unified at the level of phenomenology, *then the neural mechanisms underlying conscious experiences should also exhibit both of these properties.* That it is in virtue of expressing both of these properties that neural mechanisms do not merely correlate with, but actually account for, core features of every conscious experience.

What does it mean for a mechanism to be both integrated and informative? Let's abstract away from the brain for a moment and consider a system that consists of a large number of interacting elements, without worrying what these elements might be. As shown in the illustration on page 58, for any such system we can define a scale with two end points. At one extreme (on the left) all the elements behave randomly and independently, like molecules in a gas. This kind of system has maximum information—maximum randomness—but shows no integration at all, because every element is independent from every other.

At the other extreme (on the right), all the elements do exactly the same thing, so that the state of each element is completely determined by the state of the other elements in the system. No randomness at all. This would be like the arrangement of atoms in a

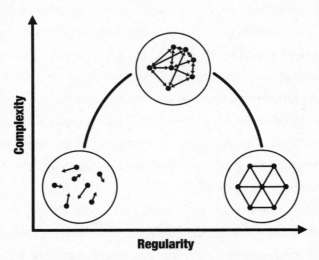

The relationship between complexity and regularity.

crystal lattice, in which the position of any individual atom is fully determined by the lattice structure, which is defined by the positions of all the other atoms. This kind of arrangement has maximum integration, but almost no information, because there are very few possible states that the system can be in.

In the middle lie systems in which individual elements may do different things, but in which there is some degree of coordination so that the system behaves to some extent "as a whole." This is the domain where both integration and information are to be found. It is also the middle ground between order and disorder, and it is here that systems are typically called "complex."

When we apply these descriptions to the brain, we can see how they shed light on the neural basis of consciousness.

In a maximally information-rich brain, all neurons would behave independently, firing randomly as if they were completely disconnected. In such a brain, measures of algorithmic complexity, like

LZW complexity, would score very high. But this brain—with lots of information but no integration—would not support any conscious states. At the other extreme, a maximally ordered brain would have all neurons doing exactly the same thing, perhaps firing in lockstep together, somewhat like what happens during global epileptic seizures. Algorithmic complexity here would be very low. This brain would also lack consciousness, but for a different reason: lots of integration, no information.

A fit-for-purpose measure of conscious level should therefore track not information itself, but rather how information and integration are jointly expressed. Such a measure—a measure of complexity in the true sense—would exemplify the real problem approach to consciousness by explicitly linking properties of mechanism to properties of experience.

As we've seen, approximations of algorithmic complexity, like LZW complexity, don't do a good job of this. They tell us a lot about information but nothing at all about integration. The PCI fares a little better. To score high on the PCI scale, the pulse of energy injected by TMS has to generate a pattern of brain activity that is difficult to compress, indicating high information. The pulse also has to travel far and wide across the cortex in order to generate an "echo"—the compressibility of which can then be assessed. However, although this cortical spread is suggestive of integration, it still falls short of what we might ideally require from such a measure. The PCI measure depends on brain activity being integrated in a rather vague way—otherwise, there would be no echo—but it doesn't measure integration in the same quantitative fashion by which information is measured. What we are looking for are measures that are directly sensitive to both integration and information, from the same data, in the same way, and at the same time.

There are several measures that do fulfill these criteria, at least in theory. Back in the 1990s, Tononi and Edelman, together with their colleague Olaf Sporns, came up with a measure they called "neural complexity," and ten years later I derived my own measure, which I called "causal density," using a different kind of mathematics. A number of newer measures, some of which we'll meet in the next chapter, have built on these foundations in increasingly sophisticated ways. All these measures attempt to quantify, in one way or another, the extent to which a system occupies the middle ground between order and disorder, where integrated information is to be found. The problem, however, is that none has yet worked particularly well when applied to actual brain imaging data.

There is something curious about this state of affairs. One might reasonably expect measures that adhere more closely to theoretical principles to perform better in practice than measures like algorithmic complexity, which are only weakly tied to the underlying theory. But this is not what we see, so what is going on? One possibility is that the theory itself is misguided. However, my intuition is that we simply need to do more work in refining the mathematics so that the measures do what we would like them to do, as well as in developing improved brain imaging methods to deliver the right kind of data for them to work on.

THE SEARCH FOR a true consciousness meter therefore goes on. It is worth emphasizing that progress so far has been considerable. It's now widely recognized that conscious level is not the same thing as wakefulness, and we already have a number of brain-based measures of conscious level that perform impressively in tracking different global states of consciousness, and in detecting residual awareness in brain-injured patients. Massimini's PCI has been particularly significant. It is both clinically useful and based firmly on sound theoretical principles of information and integration, so that it effectively bridges neural mechanisms and universal properties of conscious experience—"real problem" style. Other measures, based on different but related principles, are emerging all the time, and easy-to-use approximations, like estimating the algorithmic complexity of spontaneous brain data, are revealing fascinating connections between conscious level and conscious content.

Yet a fundamental question still remains. Is consciousness more like *temperature*—reducible to and identifiable with a basic property of the physical (or informational) universe? Or is it more like *life*, a constellation of many different properties, each with its own explanation in terms of underlying mechanisms? The approaches to measuring consciousness we've met up to now take their cue from the temperature story, but my intuition is that in the end, they may fit better with the analogy from life. For me, "integration" and "information" are general properties of most—perhaps all—conscious experiences. But this doesn't mean that consciousness *is* integrated information in the same way that temperature *is* mean molecular kinetic energy.

To see what drives this intuition, we need to push the analogy between consciousness and temperature as far as it can go, to see whether and when it breaks down. It's time to meet the "integrated information theory" of consciousness.

3

PHI

It's July 2006 and I'm in Las Vegas, eating gelato with Giulio Tononi. We're at the Venetian hotel, and I have almost no idea what's going on. I'd flown in the day before from London, and deep inside the Venetian it's always early evening, the fake stars just beginning to glimmer against the fake azure sky, the fake gondolas drifting past the fake palazzos. They keep it that way so that people stay there spending their money, in a permanent state of *aperitivo*, unaware how much time has passed. I'm jetlagged, slightly drunk—we've moved on here after a long dinner—and for several hours now we've been debating the details of the hugely ambitious "integrated information theory" of consciousness—or IIT. This is Tononi's brainchild, and more so than any other neuroscientifically motivated theory, it tackles the hard problem of consciousness head-on. IIT says that subjective experience is a property of patterns of cause and effect, that information is as real as mass or energy, and that even atoms may be a little bit conscious.

It's not an even-handed discussion. I spend much of the time defending a recent paper of mine that criticizes an early version of his theory. Giulio is gently but persistently trying to explain why

I'm wrong. I'm not sure whether it's the jet lag, the wine, or Giulio's relentless logic, but I am less sure of myself than I'd been on the flight over. The next morning, I resolve to think harder, to understand more, to prepare better, and to drink less.

I found IIT fascinating then, and continue to find it fascinating now, because it exemplifies the analogy between consciousness and temperature. According to IIT, consciousness simply *is* integrated information. In making this case, the theory upends deeply held intuitions about how mind and matter relate, and about how consciousness is woven into the fabric of the universe.

Back in 2006, IIT was not well-known. Today, it's one of the highest-profile but also most hotly debated theories in consciousness science. It's been eulogized by some of the biggest names in the field, besides Tononi himself. Christof Koch, former champion of the NCC approach, called it "a gigantic step in the final resolution of the ancient mind-body problem." But its ambition and prominence have also provoked considerable pushback. One reason for this pushback is that its approach is deeply mathematical and unapologetically complex. Of course, this is not necessarily a bad thing: nobody said that solving consciousness should be simple. Another objection is that the claims it makes are so counterintuitive that the theory must be wrong. This, too, is a dangerous intuition to rely on when faced with a phenomenon as perplexing as consciousness.

For me, the bigger problem is that IIT's extraordinary claims require extraordinary evidence, yet it is precisely the ambition of IIT—to solve the hard problem—that renders its most distinctive claims untestable in practice. The extraordinary evidence that is needed cannot be obtained. Fortunately, not all is lost. As I will explain, some predictions of IIT may be testable, at least in principle. And there are alternative interpretations of IIT, more closely aligned with the real problem than with the hard problem, which are driving the development of new measures of conscious level that are both theoretically principled *and* practically applicable.

AS ITS NAME GIVES AWAY, the concepts of "information" and "integration" are at the heart of IIT. The theory builds on ideas about measuring conscious level that we encountered in the previous chapter, but it does so in truly distinctive ways.

At the core of IIT is a single measure called "Φ" (the Greek letter phi, pronounced *fy*). The easiest way to think about Φ is that it measures how much a system is "more than the sum" of its parts, in terms of information. How can a system be more than the sum of its parts? A flock of birds provides a loose analogy: the flock seems to be more than the sum of the birds that make it up—it seems to have a "life of its own." IIT takes this idea and translates it into the domain of information. In IIT, Φ measures the amount of information a system generates "as a whole," over and above the amount of information generated by its parts independently. This underpins the main claim of the theory, which is that *a system is conscious to the extent that its whole generates more information than its parts.*

Notice that this is not merely a claim about a correlation, nor is it a real problem–style proposal about how mechanistic properties of a system account for properties of phenomenology. It is a claim about identity. According to IIT, the level of Φ is intrinsic to a system (meaning that it does not depend on an external observer), and it is identical to the amount of consciousness associated with that system. High Φ, lots of consciousness. Zero Φ, no consciousness. This is why IIT is the ultimate expression of a temperature-based view of consciousness.

What does it take to have high Φ? Although the core idea should be familiar from the previous chapter, there are some important differences, so it's worth taking it from the top.

Imagine a network of simplified artificial "neurons," each of which can be "on" or "off." To have high Φ, the network has to satisfy two main conditions. First, the global state of the network—the network "as a whole"—has to rule out a large number of alternative possible global states. This is *information*, and it reflects the observation from phenomenology that every conscious experience rules out a great many alternative possible conscious experiences. Second, there must be more information when considering the system as a whole than when dividing it into its parts (its individual neurons, or groups of neurons) and considering all the parts separately. This is *integration*, and it reflects the observation that all conscious experiences are unified, that they are experienced "all of a piece." Φ is a way of putting a number to a system that measures how high it scores on both these dimensions.

There are many ways a system can fail to have high Φ. One is to score low on information. A minimal example is a single photodiode—a simple light sensor which can be either "on" or "off." This has low or zero Φ because its state at any time carries very little information about anything. Whatever state it is in (one or zero, on or off) only ever rules out one alternative (zero or one). A single photodiode conveys at most one "bit" of information.*

Another way a system can have low Φ is to score low on integration. Imagine a large array of photodiodes, perhaps like the sensor in your phone's camera. The global state of the system is the state of the entire array, and this can carry a great deal of information. A

---

* In information theory, the "bit" is the fundamental unit of information.

sufficiently large sensor array will enter a different global state for every different state of the world it encounters, which is why cameras are useful. But this global information does not matter to the sensor itself. The individual photodiodes in the sensor are causally independent from one another—their state depends only on the level of light they each encounter. Chop the sensor into a bunch of (causally independent) photodiodes and it will work just as well. The information conveyed by the sensor array as a whole is not more than that conveyed by all the sensors, all the photodiodes, independently. This means that the information it generates is no more than the sum of its parts and so its Φ will also be zero.

Another instructive zero Φ example is a so-called "split brain" situation. Imagine a network divided into two completely separate halves. Each half of this network may have a nonzero Φ, but the network as a whole will have zero Φ. This is because there is a way of dividing the network into parts—the two halves—for which the whole is no more than the sum of the parts. This example emphasizes how Φ depends on the optimal way in which a system can be "cut up" to minimize the difference between what the whole does and what the parts do. This is one of the distinctive aspects of IIT which sets it apart from the measures of complexity described in the previous chapter.

This example also implies that a *real* split brain—following surgery to divide the cortical hemispheres, as happens in some cases of otherwise untreatable epilepsy—might harbor two independent "consciousnesses," but that there will be no single conscious entity spanning both hemispheres. In the same way, you and I are both conscious, but there is no collective conscious entity spanning the two of us, because we can be informationally split right down the middle.

Sticking with real brains for a moment, IIT accounts neatly for a number of observations about conscious level. Remember from the previous chapter that the cerebellum doesn't seem much involved in consciousness, despite containing about three-quarters of all neurons in the brain. This is explained by IIT because the anatomy of the cerebellum is comparable to the sensor array in your camera—a vast number of Φ-unfriendly semi-independent circuits. The cerebral cortex, by contrast, is packed with densely interconnected wiring that is likely to be associated with high Φ. So then why does consciousness fade during dreamless sleep, anesthesia, and coma, given that this wiring doesn't change? IIT says that, in these states, the ability of cortical neurons to interact with others is compromised in ways such that Φ vanishes.

IIT is an "axiomatic" approach to consciousness. It starts with theoretical principles rather than with experimental data. Axioms, in logic, are statements that are self-evidently true, in the sense that they are generally agreed to require no additional justification. "Two shapes that fill exactly the same space are the same shape"—from the Greek philosopher Euclid—is a good example. IIT proposes axioms about consciousness—primarily that conscious experiences are both integrated and informative—and uses these axioms to support claims about what properties the mechanisms that underlie these experiences must have. On IIT, *any* mechanism that has these properties, whether a brain or not, whether biological or not, will have nonzero Φ, and will have consciousness.

So MUCH FOR PRINCIPLES. Like any theory, IIT will stand or fall on whether its predictions are testable. The primary claim of the theory is that the level of consciousness for a system is given by its Φ.

Testing this requires measuring Φ for real systems, and this is where the trouble starts. It turns out that measuring Φ is extremely challenging and in most cases nearly or actually impossible. The main reason for this is that IIT treats "information" in an unusual way.

The standard use of information in mathematics, developed by Claude Shannon in the 1950s, is *observer-relative*. Observer-relative (or *extrinsic*) information is the degree to which uncertainty is reduced, from the perspective of an observer, by observing a system in a particular state. For example, imagine rolling a single die many times. Each time, you observe one outcome out of six possibilities: each time, five alternatives are ruled out. This corresponds to a reduction in uncertainty of a particular amount (measured in bits), and is information that is "for" the observer.

To measure observer-relative information, it's usually sufficient to observe how a system behaves, over some period of time. With dice, you can just write down what you get each time you make a new throw, and this will allow you to calculate how much information is generated by throwing any particular number. If the system is a network of neurons, it's enough to record the activity of the neurons over time. An external observer can record all the different states the neurons enter into, calculate the probabilities associated with each state, and then measure the reduction in uncertainty associated with the network being in any one of these states.

For IIT, however, information cannot be treated in this observer-relative way. This is because in IIT information—*integrated* information, Φ—actually *is* consciousness, and so if we treat information as observer-relative, then this would mean that consciousness is also observer-relative. But it isn't observer-relative. Whether I am conscious or not should not and does not depend on how you or anyone else measures my brain.

Information in IIT must therefore be treated as *intrinsic* to a system, not as relative to an external observer. It must be defined in a way that does not depend on any external observer. It must be information "for" the system itself—not for anyone or anything else. If not, the identity relation between Φ and consciousness at the core of IIT cannot hold.

In order to measure intrinsic information, it is not enough merely to observe how the system behaves over time. You—as the scientist, the external observer—have to know all the different ways a system *could* behave, even if it never actually behaves in all these ways. The distinction is between knowing what a system *actually does* over time (which is easy, at least in principle, and is observer-relative) and knowing what a system *could do* even if it never does it (which is usually difficult, if not impossible, but is observer-independent).

In the language of information theory, the difference between these situations is the difference between the "empirical" distribution of the states of a system and its "maximum entropy" distribution (the latter has its name because it reflects the maximum level of uncertainty about a system). Imagine rolling two dice several times. Perhaps you throw a seven, an eight, and an eleven, and a few other numbers, but never a twelve. In this situation, the empirical distribution would not contain a twelve, but the maximum entropy distribution would, because a twelve *could have happened*, even though in this particular sequence of throws it didn't. This means that any particular outcome—whether it was a seven, an eight, or an eleven—would generate more information (reduce more uncertainty, rule out more alternatives) with respect to the maximum entropy distribution (which includes twelve), than with respect to the empirical distribution (which doesn't).

Compared to measuring the empirical distribution of a system by just observing it over time, measuring the maximum entropy distribution is in general a very difficult thing to do. There are two ways one might go about it. The first is to perturb the system in all possible ways and see what happens, just as a child might push all the buttons on a new toy to see all the things it can do. The second is to infer the maximum entropy distribution from an exhaustive, complete knowledge of the system's physical mechanism—its "cause-effect structure." If you know everything about a mechanism, it is sometimes possible to know all the things it could do, even if in practice it doesn't do them. If I know that a die has six sides, I can figure out that two dice can generate all the numbers from two to twelve without having to make a single throw.

Unfortunately, often all we have access to are the dynamics of a system, to what a system *does*, rather than to what it *could do*. This is certainly true for brains. I can record what your brain does, at varying levels of detail, but I have no way of knowing its complete physical structure, nor can I perturb its activity in all possible ways. For these reasons, the most distinctive claim of IIT—that Φ actually *is* consciousness—is also the least testable.

THERE ARE OTHER CHALLENGES faced by attempts to measure Φ, whichever flavor of information you choose. One is that measuring any Φ requires finding the appropriate way of dividing up the system so as to best compare the "whole" with the "parts." For some systems—like split brains—it's pretty easy (split it down the middle), but in general this is a very difficult problem, since the number of possible ways of dividing up a system grows exponentially with its size.

Then there are even more fundamental questions about what counts as a system in the first place. What is the right granularity of space and time over which to calculate Φ? Is it neurons and milliseconds, or atoms and femtoseconds? Could an entire country be conscious—and would one country be more conscious than another? Could we even consider the interactions of tectonic plates over geological timescales as integrating information on a planetary scale?

It's important to recognize that these challenges—including that of measuring intrinsic information, rather than observer-relative, extrinsic information—are problems only for us as scientists, as external observers, trying to calculate Φ. According to IIT, any particular system would just *have* a Φ. It would go about its business integrating information in just the same way that when you throw a stone, it traces an arc through the sky without needing to calculate its trajectory according to the laws of gravity. Just because a theory is difficult to test doesn't mean it's wrong. It just means it's difficult to test.

Let's set aside the challenge of measuring Φ and ask what is implied by IIT, if the theory is right. It turns out that following IIT all the way through leads to some very weird consequences.

Imagine I open up your skull and glom on to your brain a fistful of new neurons, each of which wires up to your existing gray matter in some specific way. Imagine further that, as you go about your day, these new neurons never actually do anything at all. No matter what happens, no matter what you do or who you see, they never fire. Your newly augmented brain appears, to all intents and purposes, just the same as the old one. But, here's the thing, your new neurons are organized in such a way that they *could* fire if only the rest of your brain encountered some particular state which it never in fact encounters.

For example, let's say these new neurons would only ever fire if you were to eat a Densuke watermelon, a rare fruit found only on the Japanese island of Hokkaido. Assuming that you never actually eat a Densuke, so that these new neurons never fire, IIT nevertheless predicts that *all* your conscious experiences would change—albeit very subtly. This is because there are now more potential states your brain could be in—the new neurons could potentially fire—so Φ must change too.

The flip side of this scenario leads to an equally strange prediction. Imagine a bunch of neurons sitting quietly deep inside your visual cortex. Despite being wired up to other neurons—and therefore potentially able to fire, given the right inputs—these neurons are doing nothing. Then, through some clever experimental intervention, they are actively prevented from firing—they become *inactivated* rather than merely *inactive*. Even though the overall activity of the brain has not changed at all, IIT would again predict a change in conscious experience, since there are now fewer potential states the brain could enter into.

Remarkably, a version of this experiment may soon be possible, thanks to the new technology of optogenetics, which allows researchers to control the activity of precisely targeted neurons with exquisite detail. Optogenetics uses genetic techniques to modify specific neurons so that they become sensitive to light at specific wavelengths. Then, by using lasers or LED arrays to shine light into the brains of genetically modified animals, these neurons can be switched on or switched off. In principle, optogenetics could be used to *inactivate already inactive* neurons, with the effects on conscious perception—if any—being assessed. This is not a simple experiment, and it does not provide a way of measuring Φ. But the prospect of testing any aspect of IIT is exciting, and I've been lucky to be involved in recent discussions—with Giulio Tononi and others—with a view to actually getting it done.

Zooming out, another weirdness of IIT is that by making the strong claim that Φ *is* consciousness, IIT also implies that *information itself* exists—that it has some definite ontological status in our universe—a status like mass/energy and electrical charge. (Ontology is the study of "what exists.") In some sense, this aligns with the so-called "it from bit" view of the physicist John Wheeler, probably the best-known advocate of the idea that everything that exists ultimately derives from information—that information is primary, and that everything else follows from it.

And this leads to a final challenging implication: *panpsychism*. So long as there is the right kind of mechanism, the right kind of cause-effect structure in a system, there will be nonzero Φ, and there will be consciousness. IIT's panpsychism is a restrained panpsychism, not the sort in which consciousness is spread out through the entire universe like a thin layer of jam. Rather, consciousness is to be found wherever integrated information—Φ—is to be found. This could be here and there, but not everywhere.

IIT IS ORIGINAL, ambitious, and intellectually exuberant. It remains the only neuroscientific theory out there that makes a serious attempt on the hard problem of consciousness. IIT is also most definitely weird, but the fact that something is weird doesn't mean it's wrong. Almost everything about modern physics is both weird and less wrong than the physics of the past. But the success of those parts of modern physics that are now established as being less wrong has everything to do with their being experimentally testable. And this is the trouble with IIT. With its audacity comes the heavy price that its primary claim—the equivalence between Φ and conscious level—may be impossible to test.

For my money, the best way forward is to retain the fundamental insight of IIT that conscious experiences are both informative and integrated, but to relinquish the conceit that Φ is to consciousness as mean molecular kinetic energy is to temperature. This realigns IIT's insights about the structure of conscious experiences with the perspective of the real problem. Adopting this view opens the way to developing alternative, practically applicable versions of Φ, measures which end up having a lot in common with the measures of complexity that we met at the end of the previous chapter.

My colleagues Adam Barrett and Pedro Mediano and I have been following this strategy for many years now. We've developed several versions of Φ that work with observer-relative information, rather than with intrinsic information. This allows us to measure Φ based on the observable behavior of a system over time, without worrying about what it could do but never does. As things currently stand, our various versions of Φ all behave rather differently, even on very simple model systems. This means there is still more to do in developing versions of Φ that work in practice, and that—we hope— gain their empirical grip because of, and not in spite of, their basis in theoretical principles. From our perspective, this means treating "integration" and "information" as general properties of conscious experiences to be explained, not as axiomatic claims about what consciousness actually "is." In other words, treating consciousness as being more like life than like temperature.

OUR JOURNEY THROUGH LEVELS of consciousness has taken us from the oblivion of anesthesia and coma, past the hinterlands of the vegetative and minimally conscious states, through the disconnected worlds of sleep and dreaming, out into the sunlight of full wakeful awareness, and even further afield, toward the strange hyperreality of psychedelia. Linking these levels is the idea that every conscious experience is both informative and integrated, inhabiting the complex middle ground between order and disorder. This core idea has given rise to new measures, such as the PCI, which are both practically useful and capable of building explanatory bridges, real problem–style, between the physical and the phenomenal. With IIT we've reached one of the most exciting and controversial frontiers of consciousness science, where audacity meets the limits of testability and where the analogy between consciousness and temperature may finally break down. And even though I'm skeptical about the larger claims of this provocative theory, I am just as keen now to see how it develops as I was all those years ago, eating gelato with Giulio Tononi.

Looking back, Las Vegas was exactly the right place to debate IIT. Is information real? Is consciousness everywhere? In Las Vegas it's hard to believe anything is real, besides the raw feel of experience itself. Even now, years later, I can imagine myself back in the perpetual early evening of the Venetian, the fake gondolas tracing out their clockwork patterns. I am certainly conscious, but what am I conscious *of*? In the Venetian, it's tempting to think that everything is a kind of hallucination.

As we're about to see, there's some unexpected truth to this peculiar thought.

II

CONTENT

4

PERCEIVING FROM THE INSIDE OUT

I OPEN MY EYES and a world appears. I'm sitting on the deck of a tumbledown wooden house, high in a cypress forest a few miles north of Santa Cruz, California. It's early morning. Looking straight out, I can see tall trees still wreathed in the cool ocean fog that rolls in every night, sending the temperature plummeting. I can't see the ground, so the deck and the trees all seem to be floating together with me in the mist. There are some old plastic chairs—I'm sitting on one—a table, and a tray arranged with coffee and bread. I can hear birdsong, some rustling around in the back—the people I'm staying with—and a distant murmur from something I can't identify. Not every morning is like this; this is a good morning. I try to persuade myself, not for the first time, that this extraordinary world is a construction of my brain, a kind of "controlled hallucination."

Whenever we are conscious, we are conscious of something, or of many things. These are the *contents* of consciousness. To understand how they come about, and what I mean by controlled hallucination, let's change our perspective. Imagine, for a moment, that you are a brain.

Really try to think about what it's like up there, sealed inside the bony vault of the skull, trying to figure out what's out there in the world. There's no light, no sound, no anything—it's completely dark and utterly silent. When trying to form perceptions, all the brain has to go on is a constant barrage of electrical signals which are only indirectly related to things out there in the world, whatever they may be. These sensory inputs don't come with labels attached ("I'm from a cup of coffee"; "I'm from a tree"). They don't even arrive with labels announcing their modality—whether they are visual, auditory, sensations of touch, or from less familiar modalities such as thermoception (sense of temperature) or proprioception (sense of body position).*

How does the brain transform these inherently ambiguous sensory signals into a coherent perceptual world full of objects, people, and places? In part II of this book, we explore the idea that the brain is a "prediction machine," and that what we see, hear, and feel is nothing more than the brain's "best guess" of the causes of its sensory inputs. Following this idea all the way through, we will see that the contents of consciousness are a kind of waking dream—a controlled hallucination—that is both more than and less than whatever the real world really is.

HERE'S A COMMONSENSE VIEW of perception. Let's call it the "how things seem" view.

There's a mind-independent reality out there, full of objects and people and places that actually have properties like color, shape,

---

* The familiar but completely wrong idea that humans have only five senses can be traced back to Aristotle's *De Anima*—"On the Soul"—written around 350 BC.

texture, and so on. Our senses act as transparent windows onto this world, detecting these objects and their features and conveying this information to the brain, whereupon complex neuronal processes read it out to form perceptions. A coffee cup out there in the world leads to a perception of a coffee cup generated within the brain. As to who or what is doing the perceiving—well, that's the "self," isn't it, the "I behind the eyes," one might say, the recipient of wave upon wave of sensory data, which uses its perceptual readouts to guide behavior, to decide what to do next. There's a cup of coffee over there. I perceive it and I pick it up. I sense, I think, and then I act.

This is an appealing picture. Patterns of thinking established over decades, maybe centuries, have accustomed us to the idea that the brain is some kind of computer perched inside the skull, processing sensory information to build an inner picture of the outside world for the benefit of the self. This picture is so familiar that it can be difficult to conceive of any reasonable alternative. Indeed, many neuroscientists and psychologists still think about perception in this way, as a process of "bottom-up" feature detection.

Here's how the bottom-up picture is supposed to work. Stimuli from the world—light waves, sound waves, molecules conveying tastes and smells, and so on—impinge on sensory organs and cause electrical impulses to flow "upward" or "inward" into the brain. These sensory signals pass through several distinct processing stages, shown by the black arrows in the image on page 82, with each stage analyzing increasingly complex features. Let's take vision as an example. Early stages might respond to features like luminance or edges, and later, deeper stages to object parts—such as eyes and ears, or wheels and wing mirrors. Still later stages would respond to whole objects, or object categories, like faces, cars, and coffee cups.

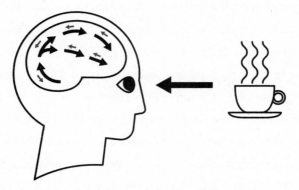

Perception as bottom-up feature detection.

In this way, the external world with its objects and people and all-sorts-of-everything becomes recapitulated in a series of features extracted from the river of sensory data flowing into the brain, like fishermen catching fish of increasing size and complexity the farther along the river they are. Signals flowing in the opposite direction— from the "top down" or the "inside out," the small gray arrows— serve only to refine or otherwise constrain the all-important bottom-up flow of sensory information.

This bottom-up view of perception fits well with what we know about the anatomy of the brain, at least at first glance. In the cortex, different perceptual modalities are associated with specific regions: visual cortex, auditory cortex, and so on. Within each region, perceptual processing is organized hierarchically. In the visual system, lower levels such as the primary visual cortex are close to sensory inputs, while higher levels, such as the infero-temporal cortex, are several stages of processing further away. In terms of connectivity, signals from each level are pooled together in the level above, so that neurons in higher levels can respond to features that may be spread out over space or time—just as one would expect.

Studies of brain activity also seem friendly to the bottom-up view. Experiments going back decades—investigating the visual systems of cats and monkeys—have repeatedly shown that neurons at early (lower) stages of visual processing respond to simple features like edges, while neurons at later (higher) stages respond to complex features like faces. More recent experiments using neuroimaging methods like fMRI have revealed much the same thing in human brains.

You can even build artificial "perceiving systems" this way—at least rudimentary ones. The computer scientist David Marr's classic 1982 computational theory of vision is both a standard reference for the bottom-up view of perception and a practical cookbook for the design and construction of artificial vision systems. More recent machine vision systems implementing artificial neural networks—such as "deep learning" networks—are nowadays achieving impressive performance levels, in some situations comparable to what humans can do. These systems, too, are frequently based on bottom-up theories.

With all these points in its favor, the bottom-up "how things seem" view of perception seems to be on pretty solid ground.

*   *   *

LUDWIG WITTGENSTEIN: "Why do people say that it was natural to think that the sun went round the Earth rather than that the Earth turned on its axis?"

ELIZABETH ANSCOMBE: "I suppose, because it looked as if the sun went round the Earth."

LUDWIG WITTGENSTEIN: "Well, what would it have looked like if it had *looked* as if the Earth turned on its axis?"

In this delightful exchange between Wittgenstein and his fellow philosopher (and biographer) Elizabeth Anscombe, the legendary Austrian thinker uses the Copernican revolution to illustrate the point that how things *seem* is not necessarily how they *are*. Although it *seems as though* the sun goes around the Earth, it is of course the Earth rotating around its own axis that gives us night and day, and it is the sun, not the Earth, that sits at the center of the solar system. Nothing new here, you might think, and you'd be right. But Wittgenstein was driving at something deeper. His real message for Anscombe was that even with a greater understanding of how things actually are, at some level things still appear the same way they always did. The sun rises in the east and sets in the west, same as always.

As with the solar system, so with perception. I open my eyes and it *seems as though* there's a real world out there. Today, I'm at home in Brighton. There are no cypress trees like there were in Santa Cruz, just the usual scatter of objects on my desk, a red chair in the corner, and beyond the window a totter of chimney pots. These objects *seem to have* specific shapes and colors, and for the ones closer at hand, smells and textures too. This is how things *seem*.

Although it may *seem as though* my senses provide transparent windows onto a mind-independent reality, and that perception is a process of "reading out" sensory data, what's really going on is—I believe—quite different. Perceptions do not come from the bottom up or the outside in, they come primarily from the top down, or the inside out. What we experience is built from the brain's predictions, or "best guesses," about the causes of sensory signals. As with the Copernican revolution, this top-down view of perception remains consistent with much of the existing evidence, leaving unchanged many aspects of how things seem, while at the same time changing everything.

This is by no means a wholly new idea. The first glimmers of a top-down theory of perception emerge in ancient Greece, with Plato's Allegory of the Cave. Prisoners, chained and facing a blank wall all their lives, see only the play of shadows cast by objects passing by a fire behind them, and they give the shadows names, because for them the shadows are what is real. The allegory is that our own conscious perceptions are just like these shadows, indirect reflections of hidden causes that we can never directly encounter.

More than a thousand years later, but still a thousand years ago, the Arab scholar Ibn al Haytham wrote that perception, in the here and now, depends on processes of "judgment and inference" rather than providing direct access to an objective reality. Hundreds of years later again, Immanuel Kant realized that the chaos of unrestricted sensory data would always remain meaningless without being given structure by preexisting conceptions, which for him included *a priori* frameworks like space and time. Kant's term "noumenon" refers to a "thing in itself"—*Ding an sich*—a mind-independent reality that will always be inaccessible to human perception, hidden behind a sensory veil.

In neuroscience the story gets going with the German physicist and physiologist Hermann von Helmholtz. In the late nineteenth century, among a string of influential contributions, Helmholtz proposed the idea of perception as a process of "unconscious inference." The contents of perception, he argued, are not given by sensory signals themselves but have to be *inferred* by combining these signals with the brain's expectations or beliefs about their causes. In calling this process "unconscious," Helmholtz understood that we are not aware of the mechanisms by which perceptual inferences happen, only of the results. Perceptual judgments—his "unconscious inferences"—keep track of their causes in the world by continually and actively updating perceptual best guesses as new sensory data arrive. Helmholtz saw himself as providing a scientific version of Kant's insight that perception cannot allow us to know things in the world directly—that we can infer only that things are there, behind the sensory veil.

Helmholtz's central idea of "perception as inference" has been remarkably influential, taking on many different forms throughout the twentieth century. In the 1950s, the "new look" movement in psychology emphasized how social and cultural factors could influence perception. For example, one widely circulated study found that children from poor families overestimated the size of coins, while those from well-to-do families didn't. Unfortunately, many experiments of this kind—while fascinating—were poorly done by today's methodological standards, so the results can't always be trusted.

In the 1970s, the psychologist Richard Gregory built on Helmholtz's ideas in a different way, with his theory of perception as a kind of neural "hypothesis-testing." According to Gregory, just as scientists test and update scientific hypotheses by obtaining data from experiments, the brain is continually formulating perceptual hypotheses about the way the world is—based on past experiences and other forms of stored information—and testing these hypotheses by acquiring data from the sensory organs. Perceptual content, for Gregory, is determined by the brain's best-supported hypotheses.

After fading in and out of the spotlight over the half century since then, the idea of perception as inference has gained new momentum in the last decade or so. A variety of new theories have blossomed under the general headings of "predictive coding" and "predictive processing." Although these theories differ in their details, they share the common proposal that perception depends on brain-based inference of some kind.

My own take on Helmholtz's enduring idea, and on its contemporary incarnations, is best captured by the notion of perception as *controlled hallucination*, a phrase I first heard from the British psychologist Chris Frith many years ago.* The essential ingredients of the controlled hallucination view, as I think of it, are as follows.

First, the brain is constantly making predictions about the causes of its sensory signals, predictions which cascade in a top-down direction through the brain's perceptual hierarchies (the gray arrows in the image below). If you happen to be looking at a coffee cup, your visual cortex will be formulating predictions about the causes of the sensory signals that originate from this coffee cup.

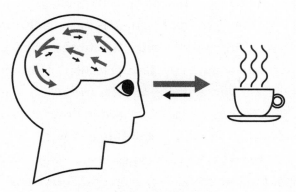

Perception as top-down inference.

Second, sensory signals—which stream into the brain from the bottom up, or outside in—keep these perceptual predictions tied in

---

* The origin of this phrase can be traced to a seminar given in the 1990s by Ramesh Jain. I have tried to trace it back further, but without success.

useful ways to their causes: in this case, a coffee cup. These signals serve as *prediction errors* registering the difference between what the brain expects and what it gets at every level of processing. By adjusting top-down predictions so as to suppress bottom-up prediction errors, the brain's perceptual best guesses maintain their grip on their causes in the world. In this view, perception happens through a continual process of *prediction error minimization*.

The third and most important ingredient in the controlled hallucination view is the claim that perceptual experience—in this case the subjective experience of "seeing a coffee cup"—is determined by the content of the (top-down) predictions, and not by the (bottom-up) sensory signals. We never experience sensory signals themselves, we only ever experience interpretations of them.

Mix these ingredients together and we've cooked up a Copernican inversion for how to think about perception. It *seems as though* the world is revealed directly to our conscious minds through our sensory organs. With this mindset, it is natural to think of perception as a process of bottom-up feature detection—a "reading" of the world around us. But what we actually perceive is a top-down, inside-out neuronal fantasy that is reined in by reality, not a transparent window onto whatever that reality may be.

And—to channel Wittgenstein once more—what would it seem like, if it *seemed as if* perception was a top-down best guess? Well, just as the sun still rises in the east and sets in the west, if it *seemed as if* perception was a controlled hallucination, the coffee cup on the table—the entirety of anyone's perceptual experience—would still seem the same way it always did, and always will.

When we think about hallucination, we typically think of some kind of internally generated perception, a seeing or a hearing of something that isn't actually there—as can happen in schizophrenia, or in psychedelic adventures like those of Albert Hofmann. These associations place hallucination in contrast to "normal" perception, which is assumed to reflect things that actually exist out in the world. On the top-down view of perception, this sharp distinction becomes a matter of degree. Both "normal" perception and "abnormal" hallucination involve internally generated predictions about the causes of sensory inputs, and both share a core set of mechanisms in the brain. The difference is that in "normal" perception, what we perceive is tied to—*controlled by*—causes in the world, whereas in the case of hallucination, our perceptions have, to some extent, lost their grip on these causes. When we hallucinate, our perceptual predictions are not properly updated in light of prediction errors.

If perception is controlled hallucination, then—equally— hallucination can be thought of as uncontrolled perception. They are different, but to ask where to draw the line is like asking where the boundary is between day and night.

LET'S TAKE THE CONTROLLED hallucination theory for a spin by asking what it means to perceptually experience *color*.

Our visual system, amazing though it is, responds to only a tiny slice of the full electromagnetic spectrum, nestled in between the lows of infrared and the highs of ultraviolet. Every color that we perceive, indeed every part of the totality of each of our visual worlds, is based on this thin slice of reality. Just knowing this is enough to tell us that perceptual experience cannot be a comprehensive representation of an external objective world. It is both less than that and more than that.

Ask a neurophysiologist and she may say that you perceive a particular color when the color-sensitive cone cells in your retina are activated in a certain proportion. This is not wrong, but it's far from the whole story. There is no one-to-one mapping between activities of color-sensitive cells and color experience. The color you experience depends on a complex interplay between the light reflected from a surface and the general illumination within the environment in which you happen to be. More precisely, it depends on how your brain makes inferences—best guesses—about how this interaction plays out.

Take a white piece of paper outdoors and it still looks white, even though the light it reflects now has a very different spectral composition, thanks to the differences between (bluish) sunlight and (yellowish) indoor light. Your visual system automatically compensates for these differences in ambient lighting—it "discounts the illuminant," as vision researchers like to say—so that your experience of color picks out an invariant property of the paper: the *way in which* the paper reflects light. The brain infers this invariant property as its best guess of the causes of its continually changing sensory inputs. Whiteness is the phenomenological aspect of this inference—it is how the brain's inferences about this invariant property appear in our conscious experience.

This means that color is not a definite property of things-in-themselves. Rather, color is a useful device that evolution has hit upon so that the brain can recognize and keep track of objects in changing lighting conditions. When I have the subjective experience of seeing the red chair in the corner of the room, this doesn't mean that the chair *actually is red*—because what could it even mean for a chair to possess a phenomenological property like redness? Chairs aren't red just as they aren't ugly or old-fashioned or avant-garde. Instead, the surface of the chair has a particular property, the *way-in-which-it-reflects-light*, that my brain keeps track of through

its mechanisms of perception. Redness is the subjective, phenomenological aspect of this process.

Does this mean that the chair's redness has moved from being
"out there" in the world to "in here" inside the brain? In one sense
the answer is clearly no. There's no red in the brain in the naive
sense of there being some kind of red pigment—or "figment"—
inside the head, to be inspected by a miniature video camera which
feeds its output into yet another visual system which itself has a
mini camera inside it . . . and so on. To assume that a perceived
property of the outside world (redness) has to be somehow re-
instantiated in the brain, in order for perception to happen, is to fall
foul of what the philosopher Daniel Dennett has called the fallacy
of "double transduction." According to this fallacy, an external
"redness" is transduced by the retina into patterns of electrical activity, which then have to be reconstituted—transduced again—into
an internal "redness." As Dennett points out, this kind of reasoning
explains nothing. The only sense in which one could locate redness
"in the brain" is simply because that's where the mechanisms underlying perceptual experience are to be found. These mechanisms are,
of course, not red.

When I look at a red chair, the redness I experience depends
both on properties of the chair and on properties of my brain. It
corresponds to the content of a set of perceptual predictions about
the ways in which a specific kind of surface reflects light. There is no
redness-as-such in the world or in the brain. As Paul Cézanne said,
"color is the place where our brain and the universe meet."

The larger claim here is that this applies far beyond the realm of
color experience. It applies to all of perception. The immersive multisensory panorama of your perceptual scene, right here and right

now, is a reaching out from the brain to the world, a writing as much as a reading. The entirety of perceptual experience is a neuronal fantasy that remains yoked to the world through a continuous making and remaking of perceptual best guesses, of controlled hallucinations.

You could even say that we're all hallucinating all the time. It's just that when we agree about our hallucinations, that's what we call reality.

LET'S CONSIDER THREE EXAMPLES of how perceptual expectations shape conscious experience, examples which you can experience for yourself.

If you were in any way connected to social media, or read a newspaper, during a particular week in February 2015, you'll remember "The Dress." On the Wednesday morning of that week, I arrived in my office to find a deluge of emails and voice mails. I'd recently co-authored a short book on visual illusions, and the media were scrambling to find explanations for a suddenly ubiquitous internet phenomenon. "The Dress" was a serendipitous photo in which a particular dress looked blue and black to some people, but white and gold to others.* Those who saw it one way were so convinced they were right, they could not believe anyone would see it differently—and so the internet erupted with claim and counterclaim.

At first, I thought it might be a hoax. To me, The Dress looked so obviously blue and black, as it did to the first four people in the lab that I showed it to, that it was both a relief and a surprise when

---

* You can find a color image of The Dress here: https://en.wikipedia.org/wiki/The_dress. What do you see?

the fifth said white and gold. So it turned out, as with the world at large, that about half the lab went for blue–black and the other half for white–gold.

An hour later I was on the BBC trying to explain what was going on. The emerging consensus was that the effect had to do with discounting the illuminant—the process of taking ambient light into account when perceiving colors. The idea was that this process might work differently for different people, in a way that normally isn't apparent and that wasn't previously known, but which just so happens to make a difference for The Dress.

People quickly pointed out that as a photo The Dress is overexposed and lacking in context—the dress itself fills most of the image—which might play tricks with how the brain generates color from context. If, for some reason, your visual system is accustomed to yellowish ambient light—perhaps you spend too much time indoors—then your visual system might be more likely to infer blue–black on the assumption of a yellowish illuminant. If, on the other hand, you are a happy healthy outdoor person, with a visual cortex frequently bathed in bluish sunlight, then maybe you'd see white–gold.

Straightaway, people started doing all sorts of experiments: staring at the photo in a dimly lit room before rushing outside into the daylight; correlating the prevalence of white–gold reports with average sunshine rates in different countries; looking at whether old people are more likely to see blue–black than are young people. Before long, a cottage industry had sprung into existence testing these and a thousand other hypotheses.

The fact that people have such different experiences and report them with such confidence, for the very same image, is compelling evidence that our perceptual experiences of the world are internal constructions, shaped by the idiosyncrasies of our personal biology and history. Most of the time, we assume that we each see the world in roughly the same way, and most of the time perhaps we do. But even if this is so, it isn't because red chairs *really are red*, it's because it takes an unusual situation like The Dress to tease apart the fine differences in how our brains settle on their perceptual best guesses.

THE SECOND EXAMPLE is a much-loved visual illusion called Adelson's Checkerboard. This example shows that the influence of predictions on perception isn't restricted to weird situations like The Dress, it happens everywhere and all the time. Take a look at the left-side checkerboard in the image on page 95, and compare squares A and B. Hopefully, A looks darker than B. It does to me, and it does to everyone I've ever shown this to. No hint of individual differences here.

In fact, A and B are precisely the same shade of gray. The checkerboard on the right proves this by joining up A and B with a rectangle that has a uniform shade of gray. Look as closely as you like, there are no changes of shading, no transitions of any kind. A and B are the same gray, though in the left-side checkerboard they persist in looking different. Knowing that they are the same doesn't help. I've stared at these images thousands of times, and A and B (on the left) stubbornly persist in seeming to be different shades of gray.*

What's going on here is that the perception of grayness is determined not by the actual light waves coming from A or B—these are

---

* When knowledge fails to affect perception, we call that perception "cognitively impenetrable."

Adelson's Checkerboard.

the same—but by the brain's best guess about what caused these particular combinations of wavelengths, and—as with The Dress— this depends on context. B is in shadow, A is not, and the brain's visual system has inscribed deep in its circuitry the knowledge that objects in shadow appear darker. In just the same way that the brain adjusts its perceptual inferences on the basis of ambient lighting, it adjusts its inferences about the shade of B on the basis of prior knowledge about shadows. This is why, in the left-side checker- board, we perceive B as being lighter than the (shadow-free) A. By contrast, for the checkerboard on the right, the shadow context is disrupted by the superimposed gray bars, so we can see that A and B are in fact identical.

This is all completely automatic. You are not—or at least were not—aware that your brain possesses and uses prior expectations about shadows when making its perceptual predictions. It's also not a failure of the visual system. A useful visual system is not meant to be a light meter, of the sort used by photographers. The function of perception, at least to a first approximation, is to figure out the most likely causes of the sensory signals, not to deliver awareness of the sensory signals themselves—whatever that might mean.

* * *

THE FINAL EXAMPLE REVEALS just how quickly new predictions can influence conscious perception. Take a look at the image below. Probably all you can see is a mess of black and white splodges. Then, after you've read the rest of this sentence, have a look at the image on page 100, before returning here.

What is this?

Good, you're back. Now have another look at the image on this page and it ought to look rather different. Where previously there was a splodgy mess, there are now distinct objects, *things* are there, and something is happening. This is a "two-tone" or "Mooney image." Once seen, it is very difficult to unsee. Two-tone images are created by taking a picture, rendering it in grayscale, and carefully thresholding it so that the details are lost in the extremes of black and white—the "two tones." If done the right way, and with the right picture, it becomes very hard to figure out what's going on. That is, until you see the original, in which case the two-tone image suddenly resolves into a coherent scene.

What's remarkable about this example is that, when you look at the original two-tone image now, the sensory signals arriving at your eyes haven't changed at all from the first time you saw it. All that's changed are your brain's predictions about the causes of this sensory data, and this changes what you consciously see.

This phenomenon is not unique to vision. There are compelling auditory examples too, which are known as "sine wave speech." Here, a spoken phrase is processed by chopping off all the high frequencies that make normal speech comprehensible. The result usually sounds like noisy whistling, making no sense at all—the auditory equivalent of a two-tone image. Then you listen to the original unprocessed speech, and then the "sine wave" version again, and suddenly all becomes clear. Just as with the two-tone images, having a strong prediction about the causes of sensory signals changes—enriches—perceptual experience.

COLLECTIVELY, THESE EXAMPLES—while admittedly and deliberately simple—reveal perception to be a generative, creative act; a proactive, context-laden interpretation of, and engagement with, sensory signals. And as I mentioned earlier, the principle that perceptual experience is built from brain-based predictions applies across the board—not only to vision and hearing, but to all of our perceptions, all of the time.

One important implication of this principle is that we never experience the world "as it is." Indeed, as Kant pointed out with his noumenon, it is difficult to know what it would mean to do so. Even something as basic as color, as we've seen, exists only in the interaction between a world and a mind. So while we might be surprised when perceptual illusions—like those we've just encountered—reveal a discrepancy between what we see (or hear, or touch) and what's there, we should be careful not to judge perceptual experiences solely in terms of their "accuracy" in directly coinciding with reality. Accurate—"veridical"—perception, understood this way, is a chimera. The controlled hallucination of our perceptual world has been designed by evolution to enhance our survival prospects, not to be a transparent window onto an external reality, a window that anyway makes no conceptual sense. In the following chapters, we will delve more deeply into these ideas, but before we do, it's worth heading off a couple of objections.

The first objection is that the controlled hallucination view of perception denies undeniable aspects of the real world. "If everything we experience is just a kind of hallucination," you might complain, "then go jump in front of a train and see what happens."

Nothing in what I say should be taken to deny the existence of things in the world, be they onrushing trains or cats or coffee cups. The "control" in controlled hallucination is just as important as the "hallucination." Describing perception this way doesn't mean that anything goes, it means that the way in which things in the world *appear* in perceptual experience is a construction of the brain.

Having said that, it is useful to distinguish between what the Enlightenment philosopher John Locke called "primary" and "secondary" qualities. Locke proposed that the primary qualities of an object are those that exist independently of an observer, such as occupying space, having solidity, and moving. An oncoming train has these primary qualities in abundance, which is why jumping in front of one is a bad idea, whether or not you are observing it, and whatever beliefs you might hold about the nature of perception. Secondary qualities are those whose existence does depend on an observer. These are properties of objects that produce sensations—or "ideas"—in the mind, and cannot be said to independently exist in the object. Color is a good example of a secondary quality, since the experience of color depends on the interaction of a particular kind of perceptual apparatus with an object.

From a controlled hallucination perspective, both primary and secondary qualities of objects can give rise to perceptual experience through an active, constructive process. In neither case, though, is the content of the perceptual experience identical to the corresponding quality of the object.

The second objection has to do with our ability to perceive new things. One might worry that we'd need a pre-formed best guess for anything that we might ever perceive, so that we are forever trapped in a perceptual world of the already expected. Imagine that you've never seen a gorilla, not in real life, on TV, or in a film—nor even in a book—and then you unexpectedly encounter one ambling along the street. I guarantee that you will now see a gorilla, a new and probably rather scary perceptual experience. In a world of the already expected, how can this happen?

The short answer is that "seeing a gorilla" is never a completely new perceptual experience. Gorillas are animals with arms and legs and fur, and you—and your ancestors—will have seen other creatures that have some or all of these features. More generally, gorillas are objects that have defined (though furry) edges, that move in reasonably predictable ways, and that reflect light in the same way that other objects of similar size, color, and texture do. The novel experience of "seeing a gorilla" is built up from perceptual predictions operating over many different levels of granularity and acquired over many different timescales—from predictions about luminance and edges to predictions about faces and posture—that together sculpt a new overall perceptual best guess, so that you see a gorilla for the first time.

The longer answer involves learning more about how the brain performs the wickedly complex neural gymnastics involved in perceptual inference—and this is precisely where we are headed in the next chapter.

This is what.

5

THE WIZARD OF ODDS

THE REVEREND THOMAS BAYES (1702–1761)—a Presbyterian minister, philosopher, and statistician who lived much of his life in Tunbridge Wells, in southern England—never got around to publishing the theorem that immortalized his name. His "Essay towards Solving a Problem in the Doctrine of Chances" was presented to the Royal Society in London two years after his death by fellow preacher-philosopher Richard Price, and much of the mathematical heavy lifting was done later on by the French mathematician Pierre-Simon Laplace. But it is Bayes whose name is forever tied to a way of reasoning called "inference to the best explanation," the insights from which are central to understanding how conscious perceptions are built from brain-based best guesses.

Bayesian reasoning is all about reasoning with probabilities. More specifically, it is about how to make optimal inferences—what we've been calling "best guesses"—under conditions of uncertainty. "Inference," a term we've encountered already, just means reaching conclusions on the basis of evidence and reason. Bayesian inference is an example of *abductive* reasoning, as distinct from *deductive* or *inductive* reasoning. Deduction means reaching conclusions by logic alone: if Jim is older than Jane, and Jane is older than Joe, then Jim is older than Joe. If the premises are true and the rules of logic are followed, deductive inferences are guaranteed to be correct. Induction involves reaching conclusions through extrapolating from a series of observations: the sun has risen in the east for all of recorded history, therefore it always rises in the east. Unlike deductive inferences, inductive inferences can be wrong: the first three balls I pulled out of the bag were green, therefore all balls in the bag are green. This may or may not be true.

Abductive reasoning—the sort formalized by Bayesian inference—is all about finding the best explanation for a set of observations, when these observations are incomplete, uncertain, or otherwise ambiguous. Like inductive reasoning, abductive reasoning can also get things wrong. In seeking the "best explanation," abductive reasoning can be thought of as reasoning backward, from observed effects to their most likely causes, rather than forward, from causes to their effects—as is the case for deduction and induction.

Here's an example. Looking out of your bedroom window one morning, you see the lawn is wet. Did it rain overnight? Perhaps, but it could also be that you forgot to turn off your garden sprinkler. The aim is to find the best explanation, or hypothesis, for what you see: *Given* the lawn is wet, what is the probability (i) that it rained overnight, or (ii) that you left the sprinkler on? In other words, we want to infer *the most likely cause for the observed data.*

Bayesian inference tells us how to do this. It provides an optimal way of updating our beliefs about something when new data come in. Bayes' rule is a mathematical recipe for going from what we already know (the *prior*) to what we should believe next (the *posterior*), based on what we are learning now (the *likelihood*). Priors, posteriors, and likelihoods are often called Bayesian "beliefs" because they represent states of knowledge rather than states of the world. (Note that a Bayesian belief is not necessarily something I-as-a-person believe. It makes equal sense to say that my visual cortex "believes" that the object in front of me is a coffee cup as to say that I believe that Neil Armstrong landed on the moon.)

Priors are the probabilities of something being the case before new data arrive. Let's say the prior probability of overnight rain is very low—perhaps you live in Las Vegas. The prior probability of having left the sprinkler on will depend on how often you use the sprinkler, and on how forgetful you are. It is also low, but not as low as the prior probability of rain.

Likelihoods, loosely speaking, are the opposite of posteriors. They formalize reasoning "forward" from causes to effects: *Given* overnight rain or overnight sprinkling, what is the probability that the lawn is wet? Like priors, these can also vary, but for now let's assume that rain and accidental sprinkling are equally likely to make for a wet lawn.

Bayes' rule combines priors and likelihoods to come up with posterior probabilities for each hypothesis. The rule itself is simple: the posterior is just the prior multiplied by the likelihood, and divided by a second prior (this is the "prior on the data"—which in this case is the prior probability of a wet lawn; we don't need to worry about this here since it is the same for each hypothesis).

Observing a wet lawn in the morning, a Good Bayesian should choose the hypothesis with the highest posterior probability—this being the most probable explanation of the data. Since in our example the prior probability of overnight rain is lower than that of accidental sprinkling, the posterior probability for rain will also be lower. A Good Bayesian will therefore choose the sprinkler hypothesis. This hypothesis is the Bayesian best guess of the causes of the observed data—it is the "inference to the best explanation."

If this mostly seems like common sense, that's because in this particular example, it is. However, there are many situations in which Bayesian inference departs from what common sense might suggest. For instance, it is easy to wrongly conclude that you have a nasty disease on the basis of a positive medical test, thanks to a common tendency to overestimate the prior probability of having rare diseases. Even if a test is 99 percent accurate, a positive result may only slightly increase the posterior probability that you have the corresponding disease, if its prevalence in the population is sufficiently low.

Let's go back to the wet lawn scenario and take it a little further. After inspecting your own lawn, you glance over at your neighbor's lawn, and you see that it, too, is wet. This is significant new information. The likelihoods for each hypothesis are now different: for the sprinkler hypothesis, only your lawn should be wet, but for the rain hypothesis, both lawns would be wet. (Likelihoods, remember, go from assumed causes to observed data.) Being a Good Bayesian, you update your posterior probabilities and find that overnight rain is now the best explanation for what you've seen—so you change your mind.

A powerful feature of Bayesian inference is that it takes the *reliability* of information into account when updating best guesses. Information that is (estimated to be) reliable should have a larger influence on Bayesian beliefs than information that is (estimated to be) unreliable. Imagine that your bedroom window is dirty, and you've lost your glasses. It looks like your neighbor's lawn might be wet, but your eyesight is so bad, and the window so grubby, that this new information is highly unreliable, and you know it. In this case, although the rain hypothesis becomes slightly more probable when you glance over the fence, the original hypothesis of accidental sprinkling might still remain in the lead.

In many situations, the process of updating Bayesian best guesses with new data happens over and over again, in an endless cycle of inference. On each iteration, the previous posterior becomes the new prior. This new prior is then used to interpret the next round of data to form a new posterior—a new best guess—and the cycle repeats. If your lawn is wet two mornings in a row, your best guess about the cause on the second day should be informed by your best guess on the first day, and so on as each new day comes along.

Bayesian inference has been applied to great benefit in all sorts of contexts, from medical diagnosis to searching for missing nuclear submarines, with new applications emerging all the time. Even the scientific method itself can be understood as a Bayesian process, in which scientific hypotheses are updated by new evidence from experiments. Conceiving of science in this way is distinct from both the "paradigm shifts" of Thomas Kuhn, in which entire scientific edifices are overturned as inconsistent evidence accumulates, and the "falsificationist" views of Karl Popper, where hypotheses are raised and tested one by one, like balloons released into the sky and then shot down. In the philosophy of science, the Bayesian perspective has most in common with the views of the Hungarian philosopher Imre Lakatos, whose analysis focuses on what makes scientific research programs work in practice, rather than on what they might ideally consist of.

A Bayesian view of science of course means that scientists' prior beliefs about the validity of their theories will influence the extent to which these theories are updated or undermined by new data. For example, I'm aware that I have a strong prior belief that brains are

Bayesian-like prediction machines. This strong belief will not only shape how I interpret experimental evidence, it will also determine the sorts of experiments that I do, to generate new evidence relevant to my beliefs. Sometimes I wonder how much evidence it would take to overturn my Bayesian belief that the brain is essentially Bayesian.

LET'S RETURN TO OUR IMAGINED BRAIN, quiet and dark inside its skull, trying to figure out what's out there in the world. We can now recognize this challenge as an ideal opportunity to invoke Bayesian inference. When the brain is making best guesses about the causes of its noisy and ambiguous sensory signals, it is following the principles of the Reverend Thomas Bayes.

Perceptual priors can be encoded at many levels of abstraction and flexibility. These range from very general and relatively fixed priors such as "light comes from above," to situation-specific priors like "the approaching furry object is a gorilla." Likelihoods in the brain encode mappings from potential causes to sensory signals. These are the "forward reasoning" components of perceptual inference, and as with priors, they can operate at many different scales of time and space. The brain continually combines these priors and likelihoods according to Bayes' rule, so that every fraction of a second a new Bayesian posterior—a perceptual best guess—is formed. And each new posterior serves as a prior for the next round of ever-changing sensory input. Perception is a rolling process, not a static snapshot.

The reliability of sensory information plays an important role here too. Unless you happen to be in a zoo, the prior probability of "gorilla" will be very low when you first catch a glimpse of something indistinctly dark and furry in the distance. Because whatever-it-is is far away, the estimated reliability of this visual input will also be low, so that your perceptual best guess is unlikely to settle immediately on "gorilla." But as the animal gets closer, visual signals become both more reliable and more informative, so that your brain's best guesses will move through a series of options—large black dog, man in gorilla suit, actual gorilla—until you confidently perceive the gorilla, hopefully still with enough time to run away.

The simplest way to think about Bayesian beliefs—priors, likelihoods, and posteriors—is as being single numbers between 0 (representing zero probability) and 1 (representing 100 percent probability). However, to understand how the reliability of sensory signals influences perceptual inference, and indeed to see how Bayes' rule might actually be implemented within the brain, we need to go a little deeper and think in terms of *probability distributions* instead.

The diagram on page 108 shows an example probability distribution for a variable X. A variable, in mathematics, is just a symbol that can take different values. A probability distribution for X describes the probability that the value of X lies within a particular range. As the diagram shows, it can be represented by a curve. The probability that X lies within a particular range is given by the area under the curve corresponding to that range. In this example, the probability that X lies between 2 and 4 is much higher than the probability that it lies between 4 and 6. As with all probability distributions, the total area under the curve sums to exactly 1. This is because when all possible outcomes are considered, something has to happen.

Probability distributions can have many different shapes. One

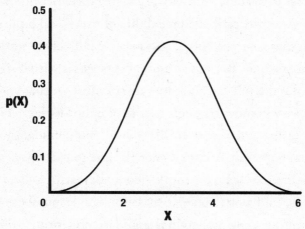

A Gaussian probability distribution.

common family of shapes, of which the present curve is an example, is the "normal," "Gaussian," or "bell curve" distribution. These distributions are fully specified by an average value or *mean* (where the curve peaks, in this case 3) and a *precision* (how spread out it is; the higher the precision, the less spread out). These quantities—mean and precision—are called the *parameters* of the distribution.*

The idea here is that Bayesian beliefs can be usefully represented by Gaussian probability distributions of this kind. Intuitively, the mean specifies the content of the belief, and the precision specifies the confidence with which the brain holds this belief. A sharply peaked (high precision) distribution corresponds to a high confidence belief. As we'll see, it is this ability to represent confidence—or reliability— that gives Bayesian inference its power.

Let's return to the gorilla example. The relevant priors, like-

---

* Sometimes the term "variance" is used instead of "precision." Precision is the inverse of variance: the higher the precision, the lower the variance.

lihoods, and posteriors can now be thought of as probability distributions, each specified by a mean and a precision. For each distribution, the mean signifies the probability of "gorilla," and the precision corresponds to the confidence the brain has in this probability estimate.

What happens when new sensory data arrive? The process of Bayesian updating is easiest to see graphically. In the diagram on page 110, the dotted curve represents the prior probability of encountering a gorilla. This curve has a low mean, indicating that gorillas are assumed to be unlikely, and a relatively high precision, indicating that this prior belief is held with high confidence. The dashed curve is the likelihood, corresponding to the sensory input. Here, the mean is higher, but the precision is lower: if a gorilla really was out there, these might be the sensory data you'd get, but you're not too confident about this. The solid curve is the posterior, representing the probability of there being a gorilla, given the sensory data. As always, this is obtained by applying Bayes' rule. When dealing with Gaussian probability distributions, applying Bayes' rule amounts to multiplying the dotted and dashed curves together, while keeping the area under the resulting curve—the posterior—limited to exactly 1.

Notice that the peak of the posterior is closer to the prior than it is to the likelihood. This is because the combination of two Gaussian distributions depends on both the means and the precisions. In this case, because the likelihood has relatively low precision—the sensory signals indicating "gorilla" are estimated to be unreliable—the posterior best guess hasn't shifted very far from the prior. However, the next moment you look, the sensory data from the gorilla may be a little clearer because it is now closer to you, and the new prior is given by the previous posterior, so the new posterior—the new best guess—will shift closer toward "gorilla." And so on until it's time to run away.

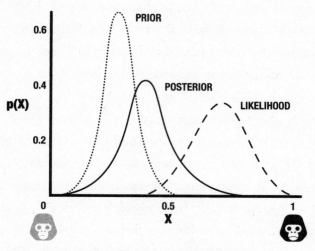

Bayesian inference with Gaussian probability distributions for
best guessing a gorilla sighting.

Bayes' theorem provides a standard of optimality for perceptual inference. It sets out best-case scenarios for what brains *should* do when trying to figure out the most likely causes of sensory inputs, whether they are gorillas or red chairs or cups of coffee. But this is only part of the story. What Bayes' theorem doesn't do is specify *how*, in terms of neural mechanisms, the brain accomplishes these feats of best guessing.

Answering this question returns us to the controlled hallucination theory of perception, and to the central claim that conscious contents are not merely shaped by perceptual predictions—they *are* these predictions.

IN THE PREVIOUS CHAPTER, I introduced the idea that perception happens through a continual process of *prediction error minimization*. According to this idea, the brain is continually generating predictions about sensory signals and comparing these predictions with the sensory signals that arrive at the eyes and the ears—and the nose, and the skin, and so on. The differences between predicted and actual sensory signals give rise to prediction errors. While perceptual predictions flow predominantly in a top-down (inside-to-outside) direction, prediction errors flow in a bottom-up (outside-to-inside) direction. These prediction error signals are used by the brain to update its predictions, ready for the next round of sensory inputs. What we perceive is given by the content of all the top-down predictions together, once sensory prediction errors have been minimized—or "explained away"—as far as possible.

The controlled hallucination view shares many features with other "predictive" theories of perception and brain function, most prominently *predictive processing*. There is, however, an important difference of emphasis. Predictive processing is a theory about the *mechanisms* by which brains accomplish perception (and cognition, and action). The controlled hallucination view, by contrast, is about how brain mechanisms explain *phenomenological properties* of conscious perception. In other words, predictive processing is a theory about how brains work, whereas the controlled hallucination view takes this theory and develops it to account for the nature of conscious experiences. Importantly, both rest on the bedrock process of prediction error minimization.

And it is prediction error minimization that provides the connection between controlled hallucinations and Bayesian inference. It takes a Bayesian claim about what the brain *should do* and turns it into a proposal about what it actually *does do*. By minimizing prediction errors everywhere and all the time, it turns out that the brain is actually implementing Bayes' rule. More precisely, it is *approximating* Bayes' rule. It is this connection that licenses the idea that perceptual content is a top-down controlled hallucination, rather than a bottom-up "readout" of sensory data.

Let's consider three core components of prediction error minimization in the brain: generative models, perceptual hierarchies, and the "precision weighting" of sensory signals.

Generative models determine the repertoire of perceivable things. In order to perceive a gorilla, my brain needs to be equipped with a generative model capable of generating the relevant sensory signals—the sensory signals that would be expected were a gorilla to be actually present. These models provide the flow of perceptual predictions which are compared against incoming sensory data to form prediction errors—which then prompt updated predictions as the brain tries to minimize these errors.

Perceptual predictions play out across many scales of space and time, so that we perceive a structured world full of objects, people, and places. A high-level prediction to see a gorilla gives rise to lower-level predictions about limbs, eyes, ears, and fur, which then cascade further down into predictions about colors, textures, and edges, and finally into anticipated variations in brightness across the visual field. These perceptual hierarchies work across the senses and even go beyond sensory data entirely. If I suddenly hear my mother's voice, my visual cortex might tune up its predictions that the approaching figure is my mother. If I know I'm in the zoo, perceptual regions of my brain will be more prepared for gorilla sightings than if I'm wandering down the street.

It's worth clarifying here that "prediction" in prediction error minimization is not necessarily about the future. It simply means going beyond the data by using a model. In statistics, the essence of prediction is in catering for the absence of sufficient data. Whether this is because predictions are about the future—one can think of the future as "insufficient data"—or about some current but incompletely known state of affairs doesn't matter.

The final key element of prediction error minimization is precision weighting. We've already seen how the relative reliability of sensory signals determines the extent to which perceptual inferences are updated. Your initial glance at a faraway gorilla, or across at your neighbor's lawn through a dirty window, will deliver sensory signals with low reliability, and so your Bayesian best guess will not shift by very much. We've also seen how reliability is captured by the precision of the corresponding probability distributions. As the diagram on page 110 showed, sensory data with low estimated precision have a weaker effect on updating prior beliefs.

I say "estimated precision" rather than simply "precision" because the precision of sensory signals is not something that is directly given to the perceiving brain. It also has to be inferred. The brain is faced not only with the challenge of figuring out the most likely causes of its sensory inputs, but also with figuring out how reliable the relevant sensory inputs are. What this means, in practice, is that the brain continually adjusts the influence of sensory signals on perceptual inference. It does this by transiently altering their estimated precision. This is what is meant by the term "precision weighting." Downweighting estimated precision means that sensory signals have less influence on updating best guesses, while up-weighting means the opposite: a stronger influence of sensory signals on perceptual inference. In this way, precision weighting plays an essential role in choreographing the delicate dance between predictions and prediction errors needed to reach a perceptual best guess.

Although this sounds complicated, we are all intimately familiar with the role of precision weighting in perception. Increasing the estimated precision of sensory signals is nothing other than "paying attention." When you pay attention to something—for example, really trying to see whether a gorilla is out there in the distance—your brain is increasing the precision weighting on the corresponding sensory signals, which is equivalent to increasing their estimated reliability, or turning up their "gain." Thinking about attention this way can explain why sometimes we don't see things, even if they are in plain view, and even if we are looking right at them. If you are paying attention to some sensory data—increasing their estimated precision—then other sensory data will have less influence on updating perceptual best guesses.

Remarkably, in some situations, unattended sensory data may have no influence at all. In 1999, the psychologist Daniel Simons developed a well-known video demonstration of this phenomenon, which he calls "inattentional blindness." If you haven't seen it, I suggest taking a look before you read on.*

Here's what happens. In the demo, Simons's subjects watch a short video in which there are two teams, each consisting of three people. One team is dressed in black, the other in white. Each team has a basketball which they pass among themselves, while wandering around in apparently random patterns. The viewer's job is to count the number of passes made only between members of the white team. This takes an effortful focus of attention, since the six players are wandering all over the place and there are two balls being passed around.

What's astonishing is that, when doing this, most people completely fail to notice a person in a black gorilla costume entering stage left, making various gorilla moves, and exiting stage right. Show them the same video again, and this time ask them to look for a

---

* www.youtube.com/watch?v=vJG698U2Mvo.

gorilla: they will immediately see it, and will often refuse to believe that it's the same video. What's happening is that focusing attention on the players in white means that the sensory signals from the players in black—and the gorilla—are afforded low estimated precision, and so have little or no influence on updating perceptual best guesses.

Something similar happened to me one afternoon many years ago, while driving to my favorite surf spot in San Diego. I'd taken a left turn where a "no left turn" sign had recently been installed; a short side road down to the ocean in the neighborhood of Del Mar. Because there was no obvious reason for this new sign, because other cars ahead of me had just made the same turn, because I'd made this turn probably hundreds of times over the years, and because I was extremely pissed off about being unfairly ticketed, I argued in a written deposition that the sign literally was not visible to me, even though it may have been visible "in principle." My defense appealed to principles of inattentional blindness. Yes, there was a new sign— but because of precision weighting in my brain, I was not able to perceive it. I took the case all the way to the California traffic court, not exactly the supreme court but far enough that my name appeared on the day's "criminal calendar." I even prepared a nice little PowerPoint presentation for the judge, which didn't help in the slightest.

Magicians, too, make use of inattentional blindness, even though they might not describe their craft in these terms. Close-up magic, in particular, involves a masterful misdirection of people's attention, so that they might not notice the queen of spades being placed behind an ear from where it later seems to appear as if from thin air. Successful pickpockets also benefit from this quirk of perceptual physiology. I once witnessed the master pickpocket Apollo Robbins effortlessly relieve some of my colleagues of their watches, wallets, and purses, a feat which was even more remarkable since many of them were experts in perception, knew all about inattentional blindness, and were fully aware of what Robbins was trying to do.

It's TEMPTING TO THINK of our interaction with the world in the following way. First, we perceive the world as it is. Then we decide what to do. Then we do it. *Sense, think, act.* This may be how things seem, but once again, how things seem is a poor guide to how they actually are. It's time to bring *action* into the picture.

Action is inseparable from perception. Perception and action are so tightly coupled that they determine and define each other. Every action alters perception by changing the incoming sensory data, and every perception is the way it is in order to help guide action. There is simply no point to perception in the absence of action. We perceive the world around us in order to act effectively within it, to achieve our goals, and—in the long run—to promote our prospects of survival. We don't perceive the world as it is, we perceive it as it is useful for us to do so.

It may even be that action comes first. Instead of picturing the brain as reaching perceptual best guesses in order to then guide behavior, we can think of brains as fundamentally in the business of generating actions, and continually calibrating these actions using sensory signals, so as to best achieve the organism's goals. This view casts the brain as an intrinsically dynamic, active system, continually probing its environment and examining the consequences.*

---

* Consider the sea squirt. In its juvenile stage this simple animal has a well-defined though rudimentary brain, which it uses as it searches for an appealing rock or lump of coral on which to spend the rest of its life filter-feeding on whatever drifts by. Having found one and attached itself, it digests its own brain, retaining only a simple nervous system. Some people have used the sea squirt as an analogy for an academic career, before and after finding a permanent university position.

In predictive processing, action and perception are two sides of the same coin. Both are underpinned by the minimization of sensory prediction errors. Until now, I've described this minimization process in terms of updating perceptual predictions, but this is not the only possibility. Prediction errors can also be quenched by performing actions in order to change the sensory data, so that the new sensory data match an existing prediction. Minimizing prediction error through action is called "active inference"—a term coined by the British neuroscientist Karl Friston.

A helpful way to think about active inference is as a kind of self-fulfilling perceptual prediction, a process by which the brain seeks out, through making actions, the sensory data that make its perceptual predictions come true. These actions can be as simple as moving one's eyes. This morning I was looking for my car keys amid the usual clutter on my desk. As my eyes were darting from place to place, not only were my moment-to-moment visual predictions being updated (empty mug, empty mug, paper clips, empty mug . . .), but my visual focus was continually interrogating the scene before me until the perceptual prediction of car keys was fulfilled.

Any kind of bodily action will change sensory data in some way, whether it's moving your eyes, walking into a different room, or tightening your stomach muscles. Even high-level "actions," like applying for a new job or deciding to get married, will cascade down into sets of bodily actions which alter sensory inputs. Every kind of action has the potential to suppress sensory prediction errors through active inference, and so every kind of action directly participates in perception.

Like all aspects of predictive processing, active inference depends on generative modeling. More specifically, active inference relies on the ability of generative models to predict the sensory consequences of actions. These are predictions of the form "If I look over there, what sensory data am I likely to encounter?" Such predictions are called "conditional" predictions—predictions about what *would* happen were something to be the case. Without conditional predictions of this kind, there would be no way for the brain to know which action, among countless possible actions, would be most likely to reduce sensory prediction errors. The actions my brain predicted as being most likely to locate my missing car keys involved visually scanning my desk, not staring out the window or waving my hands in the air.

As well as fulfilling existing perceptual predictions, active inference can also help improve these predictions. Over short timescales, actions can harvest new sensory data to help make a better best guess, or to decide between competing perceptual hypotheses. We saw an example of this at the top of this chapter, where the competing hypotheses of "overnight rain" and "accidentally left the sprinkler on" could be better discriminated by peeking over the fence at your neighbor's lawn. Another example would be tidying away all the mugs on my desk in order to help find my car keys. In each case, choosing the relevant action relies on having a generative model able to predict how sensory data would change as a result of that action.

In the long run, actions are fundamental to *learning*—which here means improving the brain's generative models by revealing more about the causes of sensory signals, and about the causal structure of the world in general. When I look over the fence to help me infer the causes of a specific wet lawn, I've also learned more about what causes wet lawns in general. In the best case, active inference can give rise to a virtuous circle in which well-chosen actions uncover useful information about the structure of the world, which is then incorporated into improved generative models, which can then enable improved perceptual inference and direct new actions predicted to deliver even more useful information.

The most counterintuitive aspect of active inference is that *action itself* can be thought of as a form of self-fulfilling perceptual prediction. In this view, actions do not merely participate in perception—actions *are* perceptions. When I move my eyes to look for my car keys, or my hands to tidy the mugs away, what's happening is that perceptual predictions about the position and movement of my body are making themselves come true.

In active inference, actions are self-fulfilling *proprioceptive* predictions. Proprioception is a form of perception which keeps track of where the body is and how it is moving, by registering sensory signals that flow from receptors situated all over the skeleton and musculature. We probably don't think much about proprioception because, in some sense, it's always there, but the simple fact that you can touch your nose with your eyes shut—try it!—demonstrates the essential role it plays in all our actions. From the perspective of active inference, touching my nose means allowing a suite of proprioceptive predictions about hand movement and position to become self-fulfilling—to overwhelm the sensory evidence that my fingers are currently *not* touching my nose. Precision weighting again plays an important role here. In order for proprioceptive predictions to make themselves come true, the prediction errors that are telling the brain where the body actually is must be attenuated, or downweighted. This can be thought of as the opposite of paying attention—a kind of "disattention" to the body, which allows it to move.

Thinking about action in this way underlines how action and perception are two sides of the same coin. Rather than perception being the input and action being the output with respect to some central "mind," action and perception are both forms of brain-based prediction. Both depend on a common process of Bayesian best guessing, on a carefully choreographed dance between perceptual predictions and sensory prediction errors, just with differences in who leads and who follows.

LET'S CHECK IN ONE FINAL time with our imagined brain, sealed inside its bony prison. We now know that this brain is far from isolated. It swims in a torrent of sensory signals from the world and the body, continually directing actions—self-fulfilling proprioceptive predictions—which proactively sculpt this sensory flow. The incoming sensory barrage is met by a cascade of top-down predictions, with prediction error signals streaming upward to stimulate ever better predictions and elicit new actions. This rolling process gives rise to an approximation to Bayesian inference, a GoodEnough Bayesianism in which the brain settles and resettles on its evolving best guess about the causes of its sensory environment, and a vivid perceptual world—a controlled hallucination—is brought into being.

Understanding controlled hallucinations this way, we now have good reasons to recognize that top-down predictions do not merely bias our perception. They *are* what we perceive. Our perceptual world alive with colors, shapes, and sounds is nothing more and nothing less than our brain's best guess of the hidden causes of its colorless, shapeless, and soundless sensory inputs.

And as we'll see next, it's not only experiences of cats, coffee cups, and gorillas that can be explained this way—but *every* aspect of our perceptual experience.

6

THE BEHOLDER'S SHARE

OUR JOURNEY INTO THE DEEP structure of perceptual experience starts with a trip to Vienna, at the turn of the twentieth century. If you'd lingered in this elegant city's cafés, salons, and opium dens during these years, you might have run into some notable characters. There was the Vienna Circle of philosophers, a group that included Kurt Gödel, Rudolf Carnap, and, from time to time, Ludwig Wittgenstein. There were the pioneers of modernist painting, Gustav Klimt, Oskar Kokoschka, and Egon Schiele, as well as the art historian Alois Riegl. And of course there was Sigmund Freud.

In the fluid intellectual atmosphere of Vienna at that time, the two cultures of art and science mingled to an unusual degree. Science wasn't placed above art, in the all too familiar sense in which art, and the human responses it evokes, are considered to be things in need of scientific explanation. Nor did art place itself beyond the reach of science. Artists and scientists—and their critics—were allies in their attempts to understand human experience in all its richness and variety. No wonder the neuroscientist Eric Kandel called this period "the age of insight," in his book of the same name.

One of the most influential ideas emerging from the age of

insight is the "beholder's share," first introduced by Riegl and later popularized by one of the major figures in twentieth-century art history, Ernst Gombrich—himself born in Vienna in 1909. Their idea highlighted the role played by the observer—the beholder—in imaginatively "completing" a work of art. The beholder's share is that part of perceptual experience that is contributed by the perceiver and which is not to be found in the artwork—or the world—itself.

The concept of the beholder's share cries out to be connected with predictive theories of perception—like the controlled hallucination theory. As Kandel put it: "The insight that the beholder's perception involves a top-down inference convinced Gombrich that there is no 'innocent eye': that is, all visual perception is based on classifying concepts and interpreting visual information. One cannot perceive that which one cannot classify."

For me, the beholder's share is particularly evident when in the company of artists like Claude Monet, Paul Cézanne, and Camille Pissarro. Standing in front of one of their Impressionist masterpieces—such as Pissarro's *Hoarfrost at Ennery*, painted in 1873 and now hanging in the Musée d'Orsay in Paris—I am drawn into a different world. One of the reasons paintings like this gain their power is because of the space they leave for the observer's visual system to perform its interpretative work. In Pissarro's painting, "palette-scrapings . . . on a dirty canvas"—as the critic Louis Leroy put it—powerfully evoke the perceptual impression of a sharply frosted field.

Impressionist landscapes attempt to remove the artist from the act of painting, to recover Gombrich's "innocent eye" by imparting to the canvas the variations in brightness that are the raw materials for perceptual inference, rather than the output of this process. To do this, the artist must develop and deploy a sophisticated understanding of how the subjective, phenomenological aspects of vision come about. Each work can be understood as an exercise in reverse engineering the human visual system, from sensory input all the way to a coherent subjective experience. The paintings become experiments into predictive perception and into the nature of the conscious experiences that these processes give rise to.

Paintings like Pissarro's are more than echoes or presentiments of a science of perception, and the beholder's share offers more than an art-historical version of prediction error minimization. What Gombrich and company bring to the table is a deep appreciation of the phenomenological, experiential nature of perception—an appreciation that is easily lost amid the nuts and bolts of priors and likelihoods and prediction errors.

"When we say the blots and brushstrokes of the Impressionist canvas 'suddenly come to life,' we mean we have been led to project a landscape into these dabs of pigment." Here, Gombrich captures something essential about conscious perception, something that applies beyond art to the nature of experience in general. When we experience the world as being "really out there," this is not a passive revealing of an objective reality, but a vivid and present projection—a reaching out to the world from the brain.

BACK IN THE LAB, efforts to unravel the ways in which perceptual expectations underpin subjective experience start with simple experiments. One experimental prediction is that we should perceive things we expect more quickly, and more easily, than things we don't expect. A few years ago, Yair Pinto—then a postdoctoral researcher with me, and now an assistant professor at the University of Amsterdam—set out to test this hypothesis, focusing as we often do on visual experience.

Yair used a setup called "continuous flash suppression" in which different images are presented to the left and right eyes. One eye is presented with a picture—in this case either a house or a face—while the other eye is presented with a rapidly changing pattern of overlapping oblongs. When the brain tries to fuse the two images into a single scene, it fails, and the changing oblongs tend to dominate, so that's what the person consciously sees. Conscious perception of the picture is suppressed by the "continuous flashing" shapes. Our experiment—illustrated on page 125—used a version of this method in which the contrast of the oblongs started high and diminished over time, while that of the picture started low and increased. This meant that after a few seconds at most, the picture—either a house or a face—became consciously visible.

To discover how perceptual expectations affected conscious perception, we cued participants with either the word "house" or the word "face" before each experimental trial. Importantly, these expectations were only partially valid. When participants were cued with the word "face," a face would be present on 70 percent of trials, but on the other 30 percent, a house would be present. The reverse was true when we led them to expect a house. By measuring how long it took for each image to emerge from flash suppression, we could determine how quickly people consciously saw a particular image—a house or a face—when it was expected, compared to when it was unexpected.

As we had predicted, people were faster and more accurate at seeing houses when houses were what they were expecting—and the same for faces. The difference in speed was small—about one-tenth of a second—but it was reliable. In our experiment, valid

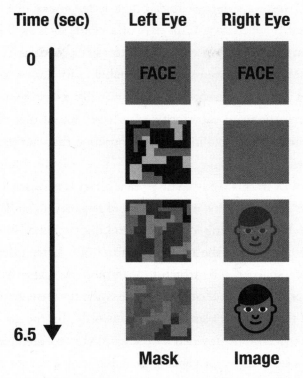

Continuous flash suppression with ramping contrast.*

perceptual expectations do indeed lead to more rapid and more ac-curate conscious perceptions.

Our study is one among a growing number that have looked into what happens when perceptual expectations are at work. In another elegant experiment, Micha Heilbron, Floris de Lange, and their col-leagues at the Donders Institute in Nijmegen took advantage of the

---

* The left eye is presented with a changing pattern of oblongs that reduces in contrast over time, while the right eye is presented with a picture (either a house or a face) that increases in contrast over time. A mirror stereoscope is used to direct each image from a computer monitor to the appropriate eye. Participants are cued to expect either a face or a house by the word "face" or "house" being presented to both eyes at the start of each trial.

so-called "word superiority effect." Single letters—like "U"—are easier to identify when they form part of a word, such as "HOUSE," than when they are part of a nonword letter string, such as "AEUVR." De Lange's team showed volunteers many examples of words and nonwords, always presented against visually noisy backgrounds. Confirming the word superiority effect, they found that individual letters were easier to read in the word condition than in the nonword condition.

The twist in this study came from a clever way of analyzing the volunteers' brain activity, which they had recorded using fMRI. By analyzing the data using a powerful technique called "brain reading," they found that the neural signature of a letter in the visual cortex was "sharper"—by which I mean more distinguishable from neural representations of other letters—when the letter formed part of a word than when it was part of a nonword. This means that perceptual expectations provided by the word context are able to alter activity at early stages of visual processing in a way that enhances perception, just as the controlled hallucination view suggests should happen.

As illuminating as experiments like these are, laboratory environments still fall far short of the richness and variety of conscious experiences out there in the wild. To get out of the lab and into the world, we need to think differently.

ONE SUMMER DAY, not so long ago and for the first time in my life, I placed a tiny amount of LSD under my tongue and lay back in the grass to see what would happen. It was a warm day with a gentle breeze and a pale blue sky scattered with wisps of cloud. After half an hour or so, just as happened to Albert Hofmann all those years before, the world started to shift and mutate. The hills and sky and clouds and sea started to pulse, becoming more vivid, deeply entrancing, entwined and interwoven with my body, somehow almost alive. Like a proper scientist, I was trying to take notes, but looking at them over the next day, I saw that my attempts had tailed off rather quickly. One memory that has stayed with me is how the clouds took on shifting but definite forms, in a way that seemed at least partially under my control. Once a particular cloud began by itself to resemble a horse—or a cat, or a person—I found that without much effort I could accentuate the effect, sometimes to absurd degrees. At one point a procession of Cilla Blacks promenaded across the horizon.*

For anyone who doubts that the brain is the organ of experience, the hallucinations produced through LSD deliver a powerful corrective. For several days afterward, I had the impression I could still see "through" my perceptual experiences, experiencing them—at least partially—as the constructions they are. I could still experience echoes of the medium, along with the perceptual message.

Of course, it's possible to see faces in clouds without a pharmaceutical boost—at least to see their hints and suggestions, their probabilistic shadows projected onto and into the sky. The general phenomenon of seeing patterns in things is called *pareidolia* (from the Greek "alongside" and "image"). For humans—and some other animals—the significance of faces means that our brains come preloaded with strong face-related prior expectations. This is why we all tend to see faces in things, to some degree, whether in clouds, pieces of toast, or even old bathroom sinks—as in the image on page 128. And because we all do it, we typically don't think of pareidolia

---

* Cilla Black was a sixties-era pop star and latter-day TV celebrity from Liverpool.

as hallucination. When a schizophrenic hears a voice commanding him to do violence to himself, or telling him that he is Jesus reborn, and when nobody else hears this voice, things are different and we call it hallucination. When, on LSD, I see Cilla Blacks marching across the sky, that, too, is hallucination.

Seeing a face in a sink.

As we now know, it would be a mistake to think of any of these phenomena—however bizarre they may appear—as wholly distinct from the normal business of perceptual best guessing. *All* our experiences, whether we label them hallucinatory or not, are always and everywhere grounded in a projection of perceptual expectations onto and into our sensory environment. What we call "hallucination" is what happens when perceptual priors are unusually strong, overwhelming the sensory data so that the brain's grip on their causes in the world starts to slide.

Inspired by this continuity between normal perception and

hallucination, in our lab we've been exploring new ways of studying how perceptual best guessing gives rise to perceptual experience, and our experiments have taken us to some strange places.

Starting from my office, if you go up two flights of stairs and wend your way through the bowels of the old chemistry department, you'll find one of our makeshift laboratory spaces—its location and purpose revealed by a piece of paper stuck to the door with Blu Tack: "VR/AR lab." Here, we use the rapidly developing technologies of virtual and augmented reality (VR/AR) to investigate the perception of the world and of the self in ways otherwise not possible. A few years ago, we decided to build a "hallucination machine" to see whether we could generate hallucination-like experiences in an experimentally controllable way, by simulating overactive perceptual priors. The project was led by Keisuke Suzuki, a senior postdoctoral researcher in the lab and our resident VR expert.

Using a 360-degree video camera, we first recorded panoramic footage of a real-world environment. We chose the main square of the university campus on Tuesday lunchtime, when students and staff mill around the weekly pop-up food market. We then processed the footage through an algorithm that Keisuke designed, which was based on Google's "deep dream" procedure, in order to generate a simulated hallucination.

The "deep dream" algorithm involves taking an artificial neural network that has been trained to recognize objects in images, and running it backward. Networks like this consist of many layers of simulated neurons, with the connections arranged so that it resembles, in some ways, the bottom-up pathway through a biological visual system. Because these networks contain only bottom-up connections, they are easy to train using standard machine learning methods. The particular network we used had been trained to identify more than a thousand different kinds of objects within images, including many different breeds of dog. It does an excellent job, even distinguishing between different varieties of husky, which all look the same to me.

The standard way these networks are used is to present them with an image and then ask what the network "thinks" is in the image. The deep dream algorithm reverses the procedure, telling the network that a particular object is present, and updating the image instead. In other words, the algorithm is projecting a perceptual prediction onto and into an image, giving it an excess of the beholder's share. For the hallucination machine, we applied this process frame by frame to the entire panoramic video, and added a few bells and whistles to cope with image continuity and so on. We replayed the deep-dreamed movie through a head-mounted display, so people could look around and experience it in an immersive way, and the hallucination machine was born.

When I first tried it out, the experience was much more compelling than I'd anticipated. While it wasn't anything like a full-blown acid trip or psychotic hallucination (as far as I know), the world was nevertheless substantially transformed. There were no Cilla Blacks, but this time dogs, and dog parts, were organically emerging out of all parts of the scene around me in a way that seemed entirely different from just pasting dog pictures onto a preexisting movie (see the image on page 131 for a black-and-white still). The power of the hallucination machine lies in its ability to simulate the effects of top-down best guesses that dogs are present, and, in doing so, to recapitulate in an exaggerated fashion the process by which we perceive and interpret visual scenes in the real world.

By programming the hallucination machine in slightly different

Still image from the hallucination machine.

ways, we can generate different kinds of simulated hallucinatory experience. For example, if we fix the activity in one of the middle layers of the network—rather than in the output layer—we end up with hallucinations of object parts, rather than of whole objects. In this case, the scene before you becomes suffused with eyes and ears and legs, a jumbled morass of dog parts pervading the entirety of your visual world. And fixing even lower layers leads to what are best described as "geometric" hallucinations, in which low-level features of the visual environment—edges, lines, textures, patterns—become unusually vivid and prominent.

The hallucination machine is an exercise in what we might call "computational phenomenology": the use of computational models to build explanatory bridges from mechanisms to properties of perceptual experience. Its immediate value lies in matching the computational architecture of predictive perception to the phenomenology of hallucination. This way, we can start to understand why specific kinds of hallucination are the way they are. But beyond this application lies the deeper and, for me, more interesting claim that by shedding light on hallucinations, we will be better able to understand normal, everyday perceptual experience as well. The hallucination machine makes clear, in a personal, immediate, and vivid way, that what we call "hallucination" is a form of uncontrolled perception. And that normal perception—in the here and now—is indeed a form of controlled hallucination.

ONE MIGHT WORRY that the controlled hallucination view is limited to explaining things like: "I see a table because that's my brain's best guess of the causes of current sensory input." (Or instead of table: face, cat, dog, red chair, brother-in-law, avocado, Cilla Black.) I think we can go much further, to account for what I like to call the "deep structure" of perception—the ways in which conscious contents appear in our experience, in time and in space and across different modalities.

Take the apparently trivial observation that our visual world is comprised largely of objects and the spaces between them. When I look at the coffee cup on my desk, in some sense I perceive its back, even though I cannot directly see this part of it. The cup appears to me to occupy a definite volume, whereas a coffee cup in a photograph or drawing does not. This is the phenomenology of "objecthood." Objecthood is a property of how visual conscious contents generally appear, rather than being a property of any single conscious experience.

Although objecthood is a pervasive feature of visual experience, it is not universal. If on a sunny day you look up at a uniform expanse of blue sky, you do not have the impression of the sky being an "object out there." And if you glance directly at the sun and then look away, the retinal afterimage seared into your vision is experienced not as an object, but as a temporary glitch. Similar distinctions apply in other modalities: people suffering from tinnitus do not experience the distressing sounds as relating to really existing things in the world, which is why it's sometimes called "ringing in the ears."

Artists have long recognized the relevance of objecthood for human perception. René Magritte's ubiquitous *The Treachery of Images* (page 134) explores the difference between an object and an image of an object. A large part of Pablo Picasso's Cubist portfolio investigates how our perception of objecthood depends on our first-person perspective. His paintings break down and rearrange objects in multiple ways, representing them from several perspectives at once. We can think of these paintings, and others like them, as exploring the principles of objecthood from the perspective of the beholder's share. Picasso's work in particular draws the observer into imaginatively creating perceptual objects out of a jumble of possibilities. As the philosopher Maurice Merleau-Ponty put it, the painter investigates through painting the means by which an object makes itself visible to our eyes.

In cognitive science, the phenomenology of objecthood has been most thoroughly explored by "sensorimotor contingency theory." According to this theory, what we experience depends on a "practical mastery" of how actions change sensory inputs. When we perceive something, the content of what we perceive is not carried by the sensory signals; instead, it emerges from the brain's implicit knowledge about how actions and sensations are coupled. In this view, vision—and all our perceptual modalities—are things an organism *does*, not passive information feeds for a centralized "mind."

In chapter 4, we described the experience of redness in terms of

René Magritte, *The Treachery of Images* (1929).
© ADAGP, Paris, and DACS, London, 2021

brain-based predictions about how surfaces reflect light. Let's now extend this explanation to objecthood. If I hold a tomato in front of me, I perceive the tomato *as having a back*, in exactly the way that doesn't happen when I look at a picture of a tomato (or a picture of a pipe, as in Magritte's painting), or a clear blue sky, or when I experience a retinal afterimage. According to sensorimotor contingency theory, I become perceptually aware of the back of the tomato, even though I cannot directly see it, because of implicit knowledge, wired into my brain, about how rotating a tomato will change incoming sensory signals.

The necessary wiring comes in the form of a generative model. As we know from the previous chapter, generative models can predict the sensory consequences of actions. These predictions are "conditional" or "counterfactual" in the sense that they are about what *could happen* or what *could have happened* to sensory signals, given some specific action. In a research paper I wrote in 2014, I proposed that the phenomenology of objecthood depends on the *richness* of these conditional or counterfactual predictions. Generative models that encode many different predictions of this kind, such as a tomato having red skin all the way around, will lead to a strong phenomenology of objecthood. But generative models that encode only a few or no such predictions, such as for a featureless blue sky or for a retinal afterimage, will lead to weak or absent objecthood.

Another situation where objecthood is typically absent is in "grapheme-color synesthesia." The term "synesthesia" refers to a kind of "mixing of the senses." People with the grapheme-color variety have experiences of color when seeing letters: for example, the letter "A" may elicit a luminous redness, regardless of its actual color on the page. Although these color experiences happen consistently and automatically—meaning that the same color is experienced whenever a particular letter is encountered, and that no conscious effort is needed for this to happen—synesthetes do not normally confuse their synesthetic colors with real "out there in the world" colors. I think this is because synesthetic colors, when compared to "real" colors, do not support a rich repertoire of sensorimotor predictions. A synesthetic "red" does not vary much as you move around, or as the ambient lighting changes, and so there won't be any phenomenology of objecthood.

In our VR lab, we've started to put some of these ideas to the test. In one recent experiment, we created a range of deliberately unfamiliar virtual objects, each defined by a variety of blobs and protrusions, which participants viewed through a head-mounted display (see page 136). We adapted the flash suppression method used in our previous face/house experiment so that each object was initially invisible but eventually broke through into consciousness.

Whereas the face/house experiment had manipulated whether a particular image was expected or not, our VR setup allowed us instead to manipulate the validity of *sensorimotor predictions* by changing the way the objects responded to actions. Participants used a joystick to rotate the virtual objects, and we could make them respond either as a real object would or by revolving in random, unpredictable directions. We predicted that normally behaving virtual objects would break through into consciousness sooner than those that violated sensorimotor predictions, and this is precisely what we found. This experiment is admittedly imperfect because it takes the speed of access to conscious perception as a proxy for the phenomenology of objecthood. But it still shows that the validity of sensorimotor predictions can affect conscious perception in specific and measurable ways.

Some virtual objects, designed to look unfamiliar.

AMONG THE MANY INTUITIVE but false ideas about perception is that changes in what we perceive correspond directly to changes in the world. But change, like objecthood, is another manifestation of the deep structure of perceptual experience. Change in perception is not simply given by change in sensory data. We perceive change through the same principles of best guessing that give rise to all other aspects of perception.

Many experiments have shown that physical change—change in the world—is neither necessary nor sufficient for the perception of change. The snake-filled image below provides a striking example in which nothing is moving, yet there can be a perceptual impression of movement, especially if you let your eyes rove around the image. What's happening is that the fine details of the picture, when seen in the periphery of your vision—out of the corner of your eye—convince your visual cortex to infer motion even though no motion is present.

"Rotating Snakes" illusion.
Akiyoshi Kitaoka*

The opposite situation, physical change without perceptual change, happens in "change blindness." This can occur when some

---

* The effect is more powerful in the original colored version—see http://www.ritsumei.ac.jp/~akitaoka/index-e.html.

aspects of an environment change very slowly, or when everything is changing at once with only some features being relevant. In one powerful video example of this phenomenon, the entire lower half of an image can change color—from red to purple—but because it happens slowly, over about forty seconds, most people don't notice the change at all, even if they are looking directly at the changing part of the image.* (This works only when people have not been primed to expect the color change. If they are actively looking for it, the change is easy to see.) This example has some similarity to inattentional blindness, which I described in the previous chapter, where people fail to see an unexpected gorilla in the midst of a basketball game. The difference here is that what people are failing to see is "change" itself.

Some people think that change blindness exposes a philosophical dilemma: After the image has changed color, are you still experiencing red (even though it's now purple), or are you now experiencing purple, in which case what were you experiencing before, given that you didn't experience any change? The resolution is to deny the premise of the question and to recognize that *perception of change* is not the same as *change of perception*. The experience of change is another perceptual inference, another variety of controlled hallucination.

And if experiences of change are perceptual inferences, so, too, are experiences of time.

TIME IS ONE of the most perplexing topics in philosophy and in physics, as well as in neuroscience. Physicists struggle to understand what it is and why it flows (if indeed it does flow), and the challenge

---

* See https://www.youtube.com/watch?v=hhXZng6o6Dk.

for neuroscientists is no less thorny. All our perceptual experiences happen in time and through time. Even our experience of the present moment seems always smeared into a relatively fixed past and a partially open future. Time flows for us too, though sometimes it crawls along, while at other times it streaks by.

We experience seconds, hours, months, and years, but we have no "time sensors" inside our brains. For vision, we have photoreceptors in the retina; for hearing, there are "hair cells" in the ear; but there is no dedicated sensory system for time. What's more, setting aside the circadian rhythm, which provides us with jet lag among other things, there is no evidence for any "neuronal clock" inside the head which measures out our experiences in time—and which would in any case be a prime example of what Daniel Dennett called a "double transduction," in which a property of the world is re-instantiated in the brain for the benefit of an assumed internal observer. Instead, like change, like all our perceptions, experiences of time are controlled hallucinations too.

Controlled by what, though? Without a dedicated sensory channel, what could provide the equivalent of sensory prediction errors? A simple and elegant solution has been proposed by my colleague, the cognitive scientist Warrick Roseboom, who joined our center in 2015 and who now leads his own research group focused on time perception. His idea is that we infer time based not on the ticking of an internal clock but on the rate of change of perceptual contents in other modalities—and he devised a clever way to test it.

Led by Warrick, our team recorded a library of videos of different durations and in different contexts—a crowded city street, an empty office, a few cows grazing in a field near the university. We then asked volunteers to watch these videos and judge how long each lasted. When they did so, they all showed characteristic biases:

underestimating the duration of long videos, and overestimating short videos. They also showed biases according to the context of each scene, rating busy scenes as lasting longer than quiet scenes, even for videos that were objectively of the same duration.

Warrick then showed the same videos to an artificial neural network that mimicked the operation of the human visual system. This network was in fact the same one we'd used in our hallucination machine. For each video, an estimate of duration was computed, based—roughly speaking—on the accumulated rate of change of activity within the network. These estimates did not involve any "inner clock" whatsoever. Remarkably, the neural network estimates and the human estimates were virtually identical, showing the same biases by duration and by context. This shows that time perception can emerge, at least in principle, from a "best guess" about the rate of change of sensory signals, without any need for an internal pacemaker.

We've recently taken this research further by looking for evidence of this process within the brain. In a study led by postdoctoral researcher Maxine Sherman, we used fMRI to record people's brain activity while they watched the same set of videos and estimated their durations. We wanted to know whether we could use activity in the visual cortex to predict how long each video seemed, as we'd been able to do in Warrick's previous study using a computational model of vision. Maxine found that we could. Brain activity in the visual system, but not in other brain regions, neatly predicted subjective duration. This is strong evidence that experiences of duration indeed emerge from perceptual best guessing, and not from the ticking of any neuronal clock.

Other experiments that might have revealed an "inner clock" have failed to do so. My favorite, by a distance, involved jumping off cranes. The neuroscientist David Eagleman set out to test the common intuition that subjective time slows down in moments of high drama, such as in the moments before a car crash. He reasoned that this subjective slowing down might be due to an internal clock running faster—more ticks of the clock in a given time period, longer perceived duration. This in turn should lead to a "speeding up" of the rate of perception since a faster clock should mean an improved ability to perceive short durations.

To test this idea, Eagleman and his team designed a special digital watch, which displayed a series of numbers that flickered so quickly that they were impossible to read in normal conditions. Then he persuaded some brave volunteers to repeatedly perform scary adrenaline-loaded leaps into the void while staring at their flickering watches. If an internal clock was indeed speeding up, then—his reasoning went—the volunteers should see the blur resolve into readable numbers while in free fall. They couldn't, so his study provided no evidence for an internal clock. Of course, absence of evidence is not evidence of absence, but still . . . what an experiment!

AT THE FRONTIER of our research into the deep structure of perceptual experience is a project investigating the perception of "reality" itself. This project, led by Alberto Mariola, a talented PhD student in the lab, involves a new experimental setup we are calling "substitutional reality." However immersive they may be, current VR environments are always distinguishable from the real world. Volunteers in our hallucination machine always know that what they are experiencing is not real, however trippy things become. Substitutional reality aims to overcome this limitation. The goal is to create a system in which people experience their environment as being real, and believe it to be real, even though it is not real.

The idea is simple. As with the hallucination machine we prerecord some panoramic video footage, but this time the footage is of the interior of the exact same VR/AR laboratory in which we conduct our experiments. When volunteers arrive in the lab, they sit on a stool in the middle of the room and put on a head-mounted display, which has a camera attached to the front. They are invited to look around the room through the camera mounted on their headset. At a certain point, we switch the feed so that the camera displays not the live real-world scene but the prerecorded video. Remarkably, most people in this situation continue to experience what they are seeing as "real," even though it no longer is.

With this setup, we can now test ideas about the conditions in which people experience their environment as being real, and—perhaps more importantly—what it takes for this normally pervasive aspect of conscious experience to break down. These situations can and do happen, not only in cases like retinal afterimages, but also in debilitating psychiatric disorders such as depersonalization and derealization, where there can be a global loss of the experienced reality of the world, and of the self.

The investigation of what makes perceptual experience seem "real" takes us all the way back to Wittgenstein's insight about the Copernican revolution: that even when we understand how things *are*—Earth going around the sun, perception as a controlled hallucination—in many ways, things will still *seem* the same way they always did. When I look at the red chair in the corner of the room, its redness (and subjective "chair"-ness) still *seem to be* really existing—veridical—properties of a mind-independent reality, rather than elaborate constructions of a best-guessing brain.

Long ago, back in the eighteenth century, David Hume made a similar observation about causality—another pervasive feature of the way we experience the world. Rather than physical causality being an objective property of the world, ready to be detected by our senses, Hume argued that we "project" causality out into the world on the basis of repeated perception of things happening in close temporal succession. We do not and we cannot directly observe "causality" in the world. Yes, things happen in the world, but what we experience as causality is a perceptual inference, in the same way that all our perceptions are projections of our brain's structured expectations onto and into our sensory environment—exercises in the beholder's share. As Hume put it, the mind has a great propensity to spread itself out into the world so that we "gild and stain" natural objects "with the colors borrowed from internal sentiment." And it's not just colors: shapes, smells, chairness, changes, durations, and causality too—all the foreground and the background features of our perceptual worlds—all are Humean projections, aspects of a controlled hallucination.

Why do we experience our perceptual constructions as being objectively real? In the controlled hallucination view, the purpose of perception is to guide action and behavior—to promote the organism's prospects of survival. We perceive the world not *as it is*, but *as it is useful for us*. It therefore makes sense that phenomenological properties—like redness, chairness, Cilla Black–ness, and causality-ness—*seem to be* objective, veridical, properties of an external existing environment. We can respond more quickly and more effectively to something happening in the world if we perceive that thing as really existing. The out-there-ness inherent in our perceptual experience of the world is, I believe, a necessary feature of a generative model that is able to anticipate its incoming sensory flow, in order to successfully guide behavior.

To put it another way, even though perceptual properties depend on top-down generative models, we do not experience the models *as* models. Rather, we perceive *with* and *through* our generative models, and in doing so, out of mere mechanism a structured world is brought forth.

AT THE START OF THIS BOOK, I promised that following the real problem approach will chip away at the hard problem of why and how any kind of physical mechanism should give rise to, correspond with, or be identical to conscious experience. Are we making progress?

We are. Starting from the principle that the brain must infer the hidden causes of its sensory inputs, we've reached a new understanding of why and how our inner universe is populated with everything from coffee cups to colors to causality-ness—things that *seem to be* properties of an external objective reality, where this *seeming-to-be* is itself a property of perceptual inference. And it is precisely the property of "seeming to be real" that adds extra fuel to dualistic intuitions about how conscious experience and the physical world relate, intuitions which in turn lead to the idea of the hard problem. It is because our perceptions have the phenomenological character of "being real" that it is extraordinarily difficult to appreciate that, in fact, perceptual experiences do not necessarily—or ever—directly correspond to things that have a mind-independent existence. A chair has a mind-independent existence; *chairness* does not.

Once we realize this, it becomes easier to recognize the hard problem as a less hard problem, or perhaps even a nonproblem. Putting it the other way around, the hard problem of consciousness seems especially hard if we interpret the contents of our perceptual experience as really existing out there in the world. Which is exactly what the phenomenology of normal conscious perception encourages us automatically to do.

As was the case with the study of life a century ago, the need to find a "special sauce" for consciousness is receding in direct proportion to our ability to distinguish different aspects of conscious experience, and to account for them in terms of their underlying mechanisms. Dissolving the hard problem is different from solving it outright, or definitively rebutting it, but it is the best way to make progress, far better than either venerating consciousness as a magical mystery or dismissing it as a metaphysically illusory nonproblem. And our mission gathers pace when we consider that it is not only experiences of the world that are perceptual constructions.

It's time to ask who, or what, is doing all this perceiving.

III

SELF

7

DELIRIUM

IN THE SUMMER OF 2014, my mother slipped into a vegetative state while in the surgical emergency unit of the John Radcliffe Hospital in Oxford. She was suffering from an undiagnosed encephalopathy—a disease of the brain. The cause was never fully identified; she'd been admitted for bowel cancer, and the neurological problems were unexpected. I hurried back from a conference I'd been attending in Brisbane, fearing the worst. She recovered, slowly, but the memory of her hour-by-hour dissolution has stayed with me. She herself remembers very little, a blessing.

Four years later, in the summer of 2018, we are in the midst of an unlikely heat wave and another World Cup summer. This time, there is no vegetative state. Instead, my mother—now eighty-three—is suffering from what I discover is called "hospital-induced delirium," a different kind of fragmentation of her sense of self and of the world around her. Two weeks previously, she'd been rushed to the Radcliffe with a recurrence of severe bowel pain. Two days after admission, while waiting to see whether the bowel problem would resolve without surgery, she developed intense hallucinations and delusions, and I drove up from Brighton to be with her.

"Delirium"—a word surfacing from the depths of the sixteenth century—comes from the Latin *delirare*, "to deviate, to be deranged." The dictionary calls it "an acutely disturbed state of mind characterized by restlessness, illusions, and incoherence." Unlike dementia, which is a chronic and degenerative condition, delirium is usually temporary. It waxes and wanes, though it can last for weeks. In my mind the word evokes Victorian asylums, so it's a surprise to hear it used diagnostically in a twenty-first-century English hospital. But on reflection, not so much a surprise, more another reminder of how far psychiatric medicine still has to travel.

For my mother the dictionary definition is accurate. When I find her on the ward, she is sitting hunched in the chair, unsmiling, disheveled, empty-eyed. She tells me about the people she has seen crawling up the walls, and she cannot remember where she is or why she is here. Her grip on reality, and on who she is, is fading.

The worst of it comes on a Friday. She trusts nobody and is convinced that there is a grand and cruel experiment being performed on her, that we—and I include myself, since in her paranoia I am frequently the ringmaster—are deliberately inducing these hallucinations through the medications we insist she takes, for malign and opaque purposes. Charming and kind in normal life, today she barks angrily at the nurses, demands to leave the hospital, tries more than once to escape, and orders the doctors to take her mad-scientist son away. This is not my mother. This is a woman who looks like my mother, but who is not my mother.

The risk factors for hospital-induced delirium include epilepsy, infection, major surgery (or a condition requiring major surgery), fever, dehydration, lack of food and lack of sleep, side effects of medication, and—importantly—unfamiliarity of place. All apply to my mother. The unfamiliarity of place is why this particular delirium is "hospital-induced."

There are few places more likely to create a dissociation from the real world than a surgical emergency unit. Constant beeps and flashing lights, hardly any signs of the outside world—a glimpse through a window if you're lucky—an entire world shrunk to a bed, a chair, and maybe a corridor. A complement of fellow travelers in varying conditions of distress and disarray, and an ever-changing parade of similar but different nurses, junior doctors, and consultants. Every day the same as the last. Delirium, though a medical emergency, is not often recognized or treated as such. Patients enter with a disease of the body and *that* is the target of treatment, not any problems of the mind or brain which may develop along the way.

Up to a third of elderly patients entering acute care develop hospital-induced delirium, and the proportion is even higher for those having surgery. Even though it usually recedes given time, there can be severe long-term consequences including reduced cognitive capacity, an increased chance of dying in the subsequent months, and heightened risk of subsequent episodes of delirium and dementia. I take a trip out to her village house, the same house I grew up in, to bring back familiar objects in the hope they provide some reorientation, some flotsam for her to cling to. A framed photograph, her spectacles, a cardigan, an old stuffed lion from my own long-ago childhood.

Delusions are rarely random. The specifics of my mother's particular delirium have a twisted logic. She believes—she *knows*—that she is the victim of a devious experiment that everyone is "in on" and in which I am complicit. Well, I *do* do experiments on people, and when I'm in the hospital I often take on a strangely fluid identity, one part son and another part doctor, trying to comfort her while at the same time going through medical notes and murmuring to the consultants and junior doctors about paraneoplastic encephalopathies and other horrid things. The brain is always trying to find closure, to make its best guess.

Some alterations in her behavior are quite subtle. Her sentences emerge with each word spoken separately, rather than as a fluent stream. "I cannot find my glasses, I do not know where they are." This state continues for many days after her primary delirium has receded. It, too, waxes and wanes. This evening saw a step backward, another regression, deflating my hopes of getting her home soon.

My own life begins to seem unreal at the edges. I am the only close family she has, so it's down to me to be here for her. Mornings and evenings are spent at the hospital, and afternoons, if I'm lucky, catching up with work, walking, and swimming in the river Thames. On these afternoons, I head to Port Meadow, a large open grassland stocked with geese, swans, cows, and wild horses. An improbable place even among the layered worlds of Oxford. The heat wave has now lasted for weeks, so that the usually muddy fields resemble an African savannah. This morning I was chased by a horse and my heart raced as I crossed the railway bridge back into the city.

It's day fourteen in the hospital for my mother, and day twelve for me. Her acute confusion has passed, though she is still changed and her condition fluctuates. Now, though, she is amazed when I tell her that she believed I was experimenting on her, that she tried to have me taken away. I hold her hand, tell her it will be OK, and hope that she will fully return to her self.

But what *is* a "self"? Is it the sort of thing that can be departed from and returned to?

The self, too, is not what it might seem to be.

8

EXPECT YOURSELF

I T MAY SEEM AS THOUGH the self—your self—is the "thing" that does the perceiving. But this is not how things are. The self is another perception, another controlled hallucination, though of a very special kind. From the sense of personal identity—like being a scientist, or a son—to experiences of having a body, and of simply "being" a body, the many and varied elements of selfhood are Bayesian best guesses, designed by evolution to keep you alive.

Let's begin our exploration of the self with a quick trip into the future. A century or so from now, teletransportation devices have been invented which can create exact replicas of any human being. Just like the machines in *Star Trek*, they work by scanning a person in exquisite detail—down to the arrangement of each individual molecule—and using the information in the scan to build a second version of that person in a distant location, for example on Mars.

After some initial apprehension, people quickly become accustomed to this technology as an efficient means of transportation. They even get used to the necessary feature that, once the replica is created, the original is immediately vaporized—a procedure that had to be built in, in order to avoid creating an explosion of identical people. From the point of view of a traveler, let's call her Eva, this poses no practical problem at all. After some reassurances from the operator, Eva simply feels that she has disappeared from place X (London) and reappeared in place Y (Mars), in an instant.

One day, there's a hitch. The vaporization module in London malfunctions and Eva—the Eva who is in London, anyway—feels like nothing's happened and that she's still in the transportation facility. A minor inconvenience. They'll have to reboot the machine and try again, or maybe leave it until the following day. But then a technician shuffles into the room, carrying a gun. He mumbles something along the lines of "Don't worry, you've been safely tele-transported to Mars, just like normal, it's just that the regulations say that we still need to . . . and, look here, you signed this consent form . . ." He slowly raises his weapon and Eva has a feeling she's never had before, that maybe this teletransportation malarkey isn't quite so straightforward after all.

The point of this thought experiment, which is called the "tele-transportation paradox," is to unearth some of the biases most of us have when we think about what it means *to be a self.*

There are two philosophical problems raised by the teletrans-portation paradox. The consciousness-in-general problem is whether we can be sure that the replica will have conscious experiences, or whether it will be a perfectly functioning equivalent but without any inner universe. I don't find this problem very interesting. If the replica is created in sufficient detail—every molecule identical!—then there's no reason to doubt it would be conscious, and conscious in exactly the same way as the original. If the replica is not com-pletely identical, then we're back to arguments about different kinds of philosophical zombie—and there's no need to go over all that again.

The more interesting problem is that of personal identity. Is the Eva on Mars (let's call her Eva$_2$) *the same person* as Eva$_1$ (the Eva still in London)? It's tempting to say, yes, she is: Eva$_2$ would feel in every way as Eva$_1$ would have felt had she actually been transported instantaneously from London to Mars. What seems to matter for this kind of personal identity is psychological continuity, not physical continuity.* But then if Eva$_1$ has not been vaporized, *which is the real Eva?*

I think the correct—but admittedly strange—answer is that *both* are the real Eva.

WE INTUITIVELY TREAT EXPERIENCES of self differently from experiences of the world. When it comes to the experience of *being you* it seems harder to resist the intuition that it reveals a genuine property of the way things are—in this case *an actual self*—rather than a collection of perceptions. One intuitive consequence of assuming the existence of an actual self is that there can be only one such self, not two, or two-thirds, or many.

The idea that the self is somehow indivisible, immutable, transcendental, *sui generis*, is baked into the Cartesian ideal of the immaterial soul and still carries a deep psychological resonance, especially in Western societies. But it has also been repeatedly held up to skeptical scrutiny by philosophers and religious practitioners, as well as more recently by psychedelic psychonauts, medical folk, and neuroscientists.

Kant, in his *Critique of Pure Reason*, argued that the concept of

---

* Even without teletransportation, the cells in our body are continuously turning over, most being replaced every ten years or so—a biological Ship of Theseus. This doesn't seem to impact our sense of personal identity very much.

the self as a "simple substance" is wrong, and Hume talked about the self as a "bundle" of perceptions. Much more recently, the German philosopher Thomas Metzinger wrote a very brilliant book called *Being No One*—a powerful deconstruction of the singular self. Buddhists have long argued that there is no such thing as a permanent self and through meditation have attempted to reach entirely selfless states of consciousness. Ayahuasca ceremonies in South America, and increasingly elsewhere, strip away people's sense of self with heady mixtures of ritual and dimethyltryptamine.

In neurology, Oliver Sacks and others have chronicled the many ways in which the self falls apart following brain disease or damage, while split-brain patients—who we met in chapter 3—raise the possibility that one self might become two. Most curious of all are craniopagus twins, who are not only physically conjoined but also share some of their brain structures. What could it mean to be an individual self, when it turns out that one craniopagus twin can feel the other drinking orange juice?

Being you is not as simple as it sounds.

BACK AT THE TELETRANSPORTATION FACILITY, Eva_1 managed to avoid the technician's murderous intent, and is coming to terms with her new situation, while Eva_2 remains blissfully unaware of the drama unfolding back on Earth.

Even though both Evas were objectively and subjectively identical at the point of replication, their identities have already started to diverge. As with identical twins setting out on their own life journeys, the process compounds inevitably over time. Even if Eva_1 had been standing right next to Eva_2, there would have been small differences in sensory inputs leading to subtle differences in behavior, and before you know it, Eva$_1$ and Eva$_2$ are experiencing different things, laying down different memories, becoming different people.

These complexities of personal identity arise for each of us in different ways. My mother's identity changed dramatically during her delirium, and though she is now recovered she seems, at least to me, both different from and recognizably the same as who she was before—just like the two Evas. The relationship between Eva$_1$ and Eva$_2$ might even be a little bit like the relationship that holds between you now and you ten years ago—or ten years in the future.

When it comes to who you are, or who I am—the me that is subjectively and objectively "Anil Seth"—things aren't as simple as at first they seem. For one thing, the sense of personal identity—the "I" behind the eyes—is only one aspect of how "being a self" appears in consciousness.

Here's how I like to break down the elements of a human self.

THERE ARE EXPERIENCES of embodied selfhood that relate directly to the body. These include feelings of identification with the particular object that happens to be your body—we feel a certain sense of ownership over our body that doesn't apply to other objects in the world. Emotions and moods are also aspects of embodied selfhood, as are states of arousal and alertness. And running below these experiences we can find deeper, formless feelings of simply being an embodied living organism—of *being a body*—without any clearly definable spatial extent or specific content. We'll come back to this bedrock layer of selfhood later. For now it's enough to think of it as the "feeling of being alive."

Moving on from the body, there's the experience of perceiving the world from a particular point of view, of having a first-person perspective—a subjective point of origin for perceptual experience which usually appears to reside inside the head, located somewhere between the eyes and slightly behind the forehead. This *perspectival self* is nowhere better illustrated than in the Austrian physicist Ernst Mach's self-portrait, also known as "View from the Left Eye."

Ernst Mach, *Self-Portrait* (1886).

Experiences of *volition*, of intending to do things—intention—and of being the cause of things that happen—agency—are also central to selfhood. This is the *volitional self*. When people talk about "free will," it is these aspects of selfhood that they are usually talking about. For many people the notion of "free will" captures that aspect of being-a-self which they're least willing to give up to science.

All these ways of being-a-self can be in place prior to any concept of personal identity—the identity that can be associated with a name, a history, and a future. As we saw with the teletransportation paradox, for personal identity to exist, there has to be a personalized prior history, a thread of autobiographical memories, a remembered past and a projected future.

This sense of personal identity, when it emerges, can be called the "narrative self". With its appearance comes the ability to experience sophisticated emotions like regret, as opposed to mere disappointment. (We humans can also suffer "anticipatory regret"—the feeling of certainty that what I'm about to do will turn out badly, that despite knowing this, I will do it anyway, and that I and others will suffer as a consequence.) Here we can see how different levels of selfhood ramify and interact—the emergence of personal identity both changes and is partly defined by the increased range of emotional states on offer.

The *social self* is all about how I perceive others perceiving me. It is the part of me that arises from my being embedded in a social network. The social self emerges gradually during childhood and continues to evolve throughout life, though it may develop differently in conditions like autism. Social selfhood brings with it its own gamut of emotional possibilities, from new ways to feel bad— like guilt and shame—to ways of feeling good, such as pride, love, and belonging.

For each of us—in normal circumstances—these diverse elements of selfhood are bound together, all of a piece, all subsumed within an overarching unified experience—the experience of *being you*. The unified character of this experience can seem so natural—as natural as the perceptual binding of color and shape when you look at a red chair—that it is easy to take for granted.

To do so would be a mistake. Just as experiences of redness are not indications of an externally existing "red," experiences of unified selfhood do not signify the existence of an "actual self." Indeed, the experience of being a unified self can come undone all too easily. The sense of personal identity, built on the narrative self, can erode or disappear entirely in dementia and in severe cases of amnesia, and it can be warped and distorted in cases of delirium, whether hospital-induced or not. The volitional self can go awry in conditions like schizophrenia and alien hand syndrome, when people experience a reduced sense of connection with their own actions, or in akinetic mutism, a disorder in which people stop interacting with their surroundings altogether. Out-of-body experiences and other dissociative disorders affect the perspectival self, while disorders of body ownership range from phantom limb syndrome—the experience of persistent, often painful sensations located in a limb that is no longer there—to somatoparaphrenia—the experience that one of your limbs belongs to someone else. In xenomelia—an extreme form of somatoparaphrenia—people experience an intense desire to amputate an arm or a leg, a drastic remedy which on rare occasions they actually carry out.

The self is not an immutable entity that lurks behind the windows of the eyes, looking out into the world and controlling the body as a pilot controls a plane. The experience of *being me*, or of *being you*, is a perception itself—or, better, a collection of perceptions—a tightly woven bundle of neurally encoded predictions geared toward keeping your body alive. And this, I believe, is all we need to be, to be who we are.

*   *   *

TAKE THE EXPERIENCE of identifying with a particular object in the world that is your body. Not only is the changeable and precariously assembled nature of these experiences evident in conditions like somatoparaphrenia and phantom limb syndrome, it can also be revealed by simple laboratory experiments. The best-known example is the "rubber hand illusion," first described more than twenty years ago, and now a cornerstone of research on embodiment.

The rubber hand illusion is easy to try out for yourself—all you need is a willing volunteer, some pieces of cardboard to form a barrier, a couple of paintbrushes, and a rubber hand. The setup is shown in the illustration on page 162. The volunteer places her (real) hand on one side of the cardboard partition, out of sight. The rubber hand is placed in front of her, in the location and orientation that her real hand would normally occupy. Then the experimenter takes the paintbrushes and gently strokes both her real hand and the rubber hand, back and forth. The idea is that when the hands are stroked synchronously, she will develop the uncanny feeling that the rubber hand is somehow actually part of her body, even though she knows that it isn't. But when the hands are stroked asynchronously, out of time, the illusion should not develop and she will not assimilate the rubber hand into her experience of what is her body.

For some people, this is an apt description of what happens, and the feeling that an evidently fake hand is somehow-but-not-entirely part of one's body is undeniably peculiar. Having said this, the actual experience varies considerably from person to person. One way to investigate this is to suddenly threaten the rubber hand with a hammer or a knife—you'll be sure to get a strong reaction when the illusion is working.

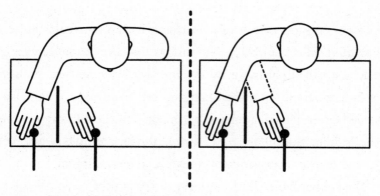

The rubber hand illusion. When the rubber hand and the real hand are stroked simultaneously (left), the experience of body ownership can shift so that the rubber hand may begin to feel like part of the person's body (right).

The rubber hand illusion fits neatly with the idea that experiences of body ownership are special kinds of controlled hallucination. The idea is that in the synchronous-stroking condition, the combination of *seeing* the rubber hand be touched and simultaneously *feeling* (but not seeing) the touch on the real hand provides enough sensory evidence for the brain to reach a perceptual best guess that the rubber hand is somehow part of the body. This happens in the synchronous—but not the asynchronous—condition because of a prior expectation that simultaneously arriving sensory signals are likely to have a common source—the rubber hand.

It is not just body parts that can be experienced differently. The whole body—and the origin of the first-person perspective—can be affected too.

In 2007, two papers appeared in the prestigious journal *Science* at almost the same time. Both described how new methods in virtual reality could be used to generate an "out-of-body-like" experience. The experiments were based on the rubber hand illusion, but now extended to the entire body. In one of the studies, conducted by a group in Lausanne led by Olaf Blanke, volunteers wore a head-mounted display through which they saw a virtual reality representation of the back of their own body from a distance of about two meters (see below). From this perspective, they then saw the virtual body being stroked with a paintbrush, again either synchronously or asynchronously with strokes applied to their own (real) body. When the stroking was synchronous, most participants reported that they felt the virtual body was, to some extent, their "own" body, and when asked to walk to where they felt their body was, they showed a drift in space toward the location of the virtual body.

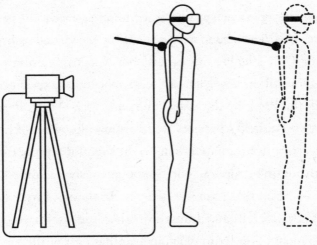

Creating a "full body illusion."

In the same way that the rubber hand illusion hints at a moment-to-moment flexibility of body ownership, experiments like this—which are called "full body illusions"—suggest that the subjective ownership of the entire body, and the location of the first-person perspective, can also be manipulated on the fly. These experiments provide fascinating evidence that experiences of "what is my body" can be dissociated, at least to some extent, from experiences of "where I am."

The idea that one's first-person perspective can leave the physical body in the form of out-of-body experiences—OBEs—is deeply inscribed in history and culture. Reports of OBEs or OBE-like experiences during traumatic near-death experiences, in operating theaters, and in the periphery of epileptic seizures have fueled beliefs in an immaterial essence-of-self. After all, if you can see yourself from the outside, then surely the basis of your consciousness must be separable from your brain?

But there's no need to reach for such dualistic skyhooks if you take the first-person perspective to be another species of perceptual inference. This view is supported not only by the virtual reality experiments of Olaf Blanke and others, but also by brain stimulation studies going all the way back to a series of seminal experiments conducted in the 1940s by the Canadian neurologist Wilder Penfield.

Among Penfield's patients was a woman known as G.A. who, when electrically stimulated in her right superior temporal gyrus—part of the brain's temporal lobe—spontaneously exclaimed, "I have a queer sensation that I am not here . . . As though I were half here and half not here." Blanke himself first became fascinated by OBEs when a patient of his, stimulated in a similar part of the brain—the angular gyrus, at the junction of the temporal and parietal lobes—reported a similar experience: "I see myself lying in bed, from above, but I only see my legs."

The common factor in cases like these is unusual activity in brain regions that deal with vestibular input (the vestibular system deals with the sense of balance) and that are also involved in multisensory integration. It seems that when normal activity in these systems becomes disrupted, the brain can reach an unusual "best guess" about the location of its first-person perspective, even while other aspects of selfhood are left unaltered.

The OBE-like experiences that sometimes accompany epileptic seizures can also be traced to disruptions in these processes. These experiences are usually divided into *autoscopic hallucinations*—in which you see your surroundings from a different perspective—and *heautoscopic hallucinations* (also known as doppelgänger hallucinations)—in which you see yourself from a different perspective. The extensive documentation of experiences like these—going back hundreds of years—is further evidence for the malleability of the first-person perspective.*

When people report apparently supernatural or otherwise bizarre experiences, like OBEs, we should take seriously their reports. They probably do have the experiences they say they have. People have had real out-of-body experiences for millennia, but this does not mean that immaterial selves or immutable souls have ever actually left any physical bodies. What these reports reveal is that first-person perspectives are put together in more complex, provisional, and precarious ways than we will ever have direct subjective access to.

IN THE VIRTUAL WORLD, the ability to alter first-person perspectives is generating some fascinating applications, many of them driven by the intriguingly named "body swap" illusion, which was described in a 2008 study led by the Swedish researcher Henrik

---

* Doppelgänger hallucinations were popularized by Fyodor Dostoevsky in his 1846 novella *The Double*. Dostoevsky was known to suffer from severe epilepsy.

Ehrsson. In the body swap setup, two people wear head-mounted displays, each with a camera attached. By swapping the camera feeds between the headsets, each person can see themselves from the other's point of view. The effect kicks in properly only when they shake hands. The idea is that seeing and simultaneously feeling the handshake provides the multisensory stimulation so that, when combined with top-down expectations, each person feels they are now somehow located in the other person's body, shaking hands with themselves. This experience puts you, albeit virtually, in the shoes of another.

I tried out virtual body swapping for myself at a small gathering in Ojai, California, in the winter of 2018. I was there along with Daanish Masood, a United Nations peace broker who also happens to be a virtual reality researcher. For several years Masood had been working closely with BeAnotherLab, the brainchild of Barcelona-based neuroscientist Mel Slater. The aim of BeAnotherLab is to adapt body swapping technology into novel "empathy generation" devices. By experiencing what it's like to perceive the world from within the virtual body of another, their idea is that empathy for the other's situation will naturally follow.

Daanish had brought his team to Ojai to demonstrate their system, called The Machine to Be Another. Their setup adds some clever choreography to the basic body swapping principle that makes the effect even more powerful. Two participants put on headsets, and first look down at their laps, so that they see their partner's body instead of their own. They then make a series of co-ordinated movements, following detailed instructions, and if they follow along closely enough, their new body will appear to respond to their commands, strengthening the experience of being the other. After some time, mirrors are held up, and each sees the mirror image of the other, as if it were himself or herself. In the final act, the curtain separating the two people is removed, and they look at themselves from within the other's body, before approaching each other and giving themselves a hug.

When it was my turn to try this out, I exchanged my own perspective with that of a rather well-to-do woman in her seventies. The experience was unexpectedly compelling. I remember looking down, flexing my (her) hand, and noticing—with some surprise— the glittering sneakers I (she) was wearing. The mirror and the final hug were particularly powerful—I was not sure whether for the experience of feeling myself to be inhabiting someone else's body, or for the experience of seeing myself from another's perspective. Only later, at dinner, did it occur to me to wonder how odd it must also have been for my partner to be suddenly transported into the first-person perspective of a mixed-race English neuroscientist with boring shoes.

I FIND IT FASCINATING that these familiar and easily taken-for-granted aspects of selfhood—subjective body ownership and the first-person perspective—can be so readily manipulated, whether with fake hands and paintbrushes or with the new technologies of virtual and augmented reality. However, there are limits to how far these manipulations can go. As I mentioned earlier, the typical experience in the rubber hand illusion is of somehow feeling the fake hand to be part of one's body, while clearly knowing that it isn't. And this "typical" experience varies considerably from person to person, with many people not feeling much at all. The same is likely true for full body and body swap illusions too.

These experimental manipulations of body ownership are in

this way very different from classic visual illusions, such as Adelson's Checkerboard, which we met in chapter 4. In the case of the checkerboard, we are so perceptually convinced that the squares are different shades of gray that we are surprised—perhaps astonished—when it is revealed that they are in fact the same shade. This kind of surprise, common in visual illusions, almost never happens in body ownership illusions. For me, the most compelling body illusion so far has been the body swapping I tried out in Ojai—but at no point was I ever close to believing I was now someone else, or somewhere else.

The subjective weakness of body ownership illusions is highlighted by a recent study which I was involved in, examining the role of hypnotic suggestibility in the rubber hand illusion. The reasoning behind this study—which was led by psychologists Peter Lush and Zoltán Dienes—was that the experimental setup of the illusion provides a strong implicit expectation for what should be experienced, and that these expectations may be enough to actually drive altered experiences of body ownership in some people. Supporting this hypothesis, we found that individual differences in the strength of the illusion correlated with how suggestible a person is, when measured on a standard scale of hypnotizability. People who were highly hypnotizable reported strong feelings of ownership (for synchronous stroking), while those who scored low on the scale were hardly susceptible at all.

On one hand, this finding fits neatly with the controlled hallucination view of body ownership, since a hypnotic suggestion can be thought of as a powerful top-down expectation—albeit one that the participant may not be aware of having. On the other hand, it poses a serious challenge for experimental research in this area, because it raises the possibility that the rubber hand illusion might be largely or entirely driven by suggestion effects. Unless studies of embodiment illusions take individual differences in suggestibility into account, which by and large they haven't, it is difficult for them to say anything specific about the mechanisms involved. This holds whether we're talking about rubber hands, out-of-body-like experiences, body swap illusions, or any other situation in which people are led—implicitly or explicitly—to expect a particular body-related experience.

There is a sharp contrast between these subjectively mild body ownership illusions and the powerfully altered experiences seen in clinical conditions like somatoparaphrenia, xenomelia, and phantom limb syndrome, or in the vivid out-of-body experiences associated with seizures or triggered by direct brain stimulation. These dramatic distortions are much more like classical visual illusions in that they elicit much greater conviction from those having the unusual experience. And for this reason, they provide much stronger evidence that experiences of embodiment and perspective are indeed constructions of the brain.

LET'S MOVE ON TO MATTERS of personal identity and to the emergence of the "narrative" and the "social" selves. As we saw with the teletransportation paradox, it's at these levels that an entity experiences itself as continuous from one moment to the next, from one day, or week, or month to the next, and—to some extent—across an entire life span. These are the levels of selfhood at which it makes sense to associate the self with a name, with memories of the past, and with plans for the future. At these levels we become aware that we have a self—we become truly *self-aware*.

These higher reaches of selfhood are fully dissociable from the embodied self. Many nonhuman animals, as well as human infants, may experience embodied selfhood without having—or missing—any accompanying sense of personal identity. And while adult humans normally experience all these forms of selfhood in an integrated and unified way, when the narrative and social aspects of self are diminished or destroyed, the impact can be devastating.

Clive Wearing is a British musicologist renowned for editing the works of the Renaissance composer Orlande de Lassus, for working as a choirmaster in London, and for reshaping the musical content of BBC Radio 3 during the early 1980s. In March 1985, at the height of his career, he suffered a devastating brain infection, a herpes encephalitis that wrought massive damage to his hippocampus in both cerebral hemispheres, producing in him one of the most profound amnesias ever documented.*

Clive has immense problems recalling old memories (retrograde amnesia) and, especially, laying down new memories (anterograde amnesia). Remarkably, he seems to exist in a permanent present of between seven and thirty seconds. He's now in his eighties, and it's likely that he still experiences his life as a continuous series of mini awakenings, as if—every twenty seconds or so—he has just emerged from a coma, or from anesthesia. His narrative self has been annihilated.

The kind of memory Clive has lost is his episodic, autobiographical memory—memory of events located in time and space (episodic), including, most important, those events involving himself (autobiographical). His diaries make for harrowing reading. They

---

* The hippocampus is a small, curved structure lying deep in the medial temporal lobe, which has long been associated with memory consolidation. The name comes from the Greek word for "seahorse."

are filled with repeated descriptions of a "first" awakening, one after the other, with previous assertions—some written just moments ago—crossed out and sometimes angrily obliterated.

> 8:31am Now I am really, completely, awake
> 9:06am Now I am perfectly, overwhelmingly awake
> 9:34am Now I am superlatively, actually awake

These diaries, and the conversations with Clive recorded by his wife, Deborah, in her book *Forever Today*, testify to the assault on his sense of his personal identity inflicted by the damage to his brain. His inability to string together a self-narrative over time means that what-it-is-to-be-him has been, for more than thirty years, a continual starting from scratch, a fleeting presence with no stable "I" around which to organize the flow of perceptions of world and self. Marooned in the present by the depths of his amnesia, his loss of a past and of a future is so dislocating that he even questions whether he is, or was, alive. Deborah Wearing writes: "Clive was under the constant impression that he had just emerged from unconsciousness because he had no evidence in his own mind of ever being awake before . . . 'I haven't heard anything, seen anything, touched anything, smelled anything,' he would say. 'It's like being dead.'"

At the same time, other aspects of Clive's sense of self remain fully intact. He has no problems with experiences of body ownership, with the origin of his first-person perspective, or even with making voluntary actions. His love for his wife remains undiminished even though sometimes he cannot remember meeting her—they married just a year before his illness. And when Clive plays the piano, or sings, or conducts, the music falls freely from him with a fluency that makes him seem whole again.

For Clive, these moments of love and music are transformational and redemptive. Oliver Sacks, in a characteristically evocative *New Yorker* piece, described his situation like this: "He no longer has any inner narrative; he is not leading a life in the sense that the rest of us do. And yet one has only to see him at the keyboard or with Deborah to feel that, at such times, he is himself again and wholly alive."

Despite these moments of grace, Clive's situation is undoubtedly tragic. The destruction of his narrative self is more than just a deficit of memory; it brings an inability to perceive himself as continuous over time, and with that comes an erosion of his fundamental sense of personal identity that most of us, quite naturally, take for granted. Memory is not the be-all and end-all of selfhood, but as this story tells us, and as many of us know through family and friends in the hinterlands of dementia or of Alzheimer's disease, the persistence and continuity of self-perception is difficult to do without.

THE POWER OF Clive and Deborah's love for each other to restore Clive's sense of identity brings us to the "social self."

Humans, like many other animals, are social creatures. Perceiving the state of mind of another is a crucial ability for social creatures in all sorts of contexts and in all manner of societies. This ability—sometimes called "theory of mind"—is often thought to develop rather slowly in humans, but it comes to play a key role for almost all of us throughout our lives.

At times we can be acutely aware of this, for example when we are worrying what a partner, friend, or colleague may be thinking about us. But even when we aren't ruminating on our social interactions, our ability to perceive others' intentions, beliefs, and desires is always operating in the background, guiding our behavior and shaping our emotions.

There is a vast literature on social perception and theory of mind, encompassing psychology, sociology, and more recently the emerging field of social neuroscience. Much of this literature examines these topics in terms of their importance for guiding social interactions. Here, I want to turn the lens inward, to consider how the experience of *being me* depends, in a substantial way, on how I perceive others perceiving me.

Social perception—perception of the mental states of others— is not just a matter of explicit reasoning or "thinking about" what others may or may not be thinking. Much of our social perception is automatic and direct. We form perceptions of others' beliefs, emotions, and intentions as naturally and effortlessly as we form perceptions of cats and coffee cups and chairs and even of our own bodies. When I pour myself another glass of wine and I see my friend has moved her empty glass closer, there's no need for me to rationally figure out what her intention is; I simply perceive that she'd like some more wine too, and that I should've poured hers first. I perceive these mental states as effortlessly, though not necessarily as accurately, as I perceive the glass itself.

How does this happen? The answer, I think, lies again with the idea of the brain as a prediction machine, and with perception as a process of inferring the causes of sensory signals.

Both nonsocial and social perception involve the brain making best guesses about the causes of sensory inputs. Sometimes, as we all know, we can get things very wrong when perceiving what's in another's mind, whereas we never confuse a wineglass with a car (unless we are hallucinating). One reason for the inherent ambiguity of social perception is that the relevant causes are more deeply hidden. The light waves that elicit perception of a wineglass originate more or less directly from the glass itself, but the sensory signals relevant to others' mental states must pass through a number of intermediate stages—through facial expressions, gestures, and speech acts—with each stage creating a new opportunity for an off-target inference.

Just as with visual perception, social perceptions depend on context and expectation, and we can try to minimize "social prediction errors" through changing sensory data—an interpersonal form of active inference—as well as by updating predictions. Active inference in social perception amounts to behaving so as to change another's mental state to bring it in line with what we predict—or desire—it to be. For example, we smile not only to express our own pleasure but also to change the way our companion is feeling, and when we speak, we are trying to insert thoughts into another's mind.

These ideas about social perception can be linked to the social self in the following way. The ability to infer others' mental states requires, as does all perceptual inference, a generative model. Generative models, as we know, are able to generate the sensory signals corresponding to a particular perceptual hypothesis. For social perception, this means a hypothesis about another's mental states. This implies a high degree of reciprocity. My best model of your mental states will include a model of how you model my mental states. In other words, I can understand what's in your mind only if I try to understand how you are perceiving the contents of my mind. It is in this way that we perceive ourselves refracted through the minds of others. This is what the social self is all about, and these socially nested predictive perceptions are an important part of the overall experience of being a human self.

One intriguing implication of this construal of the social self is that self-awareness—the higher reaches of selfhood comprising both narrative and social aspects—might necessarily require a social context. If you exist in a world without any other minds—more specifically, without any other *relevant* minds—there would be no need for your brain to predict the mental states of others, and therefore no need for it to infer that its own experiences and actions belong to any self at all. John Donne's seventeenth-century meditation that "no man is an island" could be literally true.

ARE YOU THE SAME PERSON you were yesterday? Perhaps a better question: Do you experience being yourself in the same way you did yesterday? Probably—barring some major overnight incident—you will say yes. What about last week, last month, last year, ten years ago, when you were four years old; or what about when you are ninety-four—will you be the same person then? Will it *seem* that way to you?

A striking but often overlooked aspect of conscious selfhood is that we generally experience ourselves as being continuous and unified across time. We can call this the *subjective stability of the self.* It holds not only in terms of a continuity of autobiographical memory, but in a deeper sense of experiencing oneself as persisting from moment to moment, whether at the level of the biological body or at the level of personal identity.

Compared to perceptual experiences of the outside world, self-related experiences are remarkably stable. Our perceptions of the world are always changing, objects and scenes coming and going in a continual flux of events. Self-related experiences seem to change much less. Even though we *know* we change over time—most of us have more than enough photographic evidence for that—it still *seems* to us that we don't change all that much. Unless we're suffering from psychiatric or neurological illness, the experience of being a self seems to be an enduring center within a changing world. William James—the nineteenth-century pioneer of psychology—said it well: "Contrary to the perception of an object, which can be perceived from different perspectives or even cease to be perceived, we experience 'the feeling of the same old body always there.'"

Now, you might think there's nothing to see here. After all, bodies—and other targets of self-related perception—plausibly do change less than the things we perceive out there in the world. I can move from room to room, but my body and my actions and my first-person perspective always accompany me. On these grounds it may not be surprising that the self is experienced as changing less than the world. But I think there's more to it than this.

As we saw in chapter 6, the experience of change is itself a perceptual inference. Our perceptions may change, but this doesn't mean that we perceive them as changing. This distinction is exemplified by the phenomenon of "change blindness," in which slowly changing things (in the world) do not evoke any corresponding experience of change. The same principle will apply to self-perceptions too. We are becoming different people all the time. Our perceptions of self are continually changing—you are a slightly different person now than when you started reading this chapter—but this does not mean that we perceive these changes.

This subjective blindness to the changing self has consequences. For one thing, it fosters the false intuition that the self is an immutable entity, rather than a bundle of perceptions. But this is not the reason that evolution designed our experiences of selfhood this way. I believe that the subjective stability of the self goes beyond even the change blindness warranted by our slowly changing bodies and brains. We live with an exaggerated, extreme form of self-change-blindness, and to understand why, we need to understand the reason we perceive ourselves in the first place.

We do not perceive ourselves in order to know ourselves, we perceive ourselves in order to control ourselves.

9

BEING A BEAST MACHINE

We do not see things as they are, we see them as we are.
ANAÏS NIN

SELF-PERCEPTION IS NOT ABOUT discovering what's out there
in the world, or in here, in the body. It's about physiological control
and regulation—it's about staying alive. To understand why this
is so, and what it means for *all* our conscious experiences, let's
begin by looking back at a very old debate about how life and mind
relate.

In the Great Chain of Being—the medieval Christian hierarchy
of all matter and all life—God is at the top. Just below are angelic beings, then humans—with various socially convenient subdivisions—
then other animals, plants, and finally minerals. Every thing in its
place and every thing having its own powers and abilities, determined by its place in the Chain.

Within the Chain, we humans are balanced awkwardly between the spiritual realm inhabited by God and the angels, and the
physical realm of animals, plants, and minerals. We possess immor-
tal souls and are capable of reason, love, and imagination, but we are

also susceptible to physical passions—like pain, hunger, and sexual desire—through being tied to a physical body.

For centuries, especially in Europe, the Great Chain of Being—or *Scala Naturae*—provided a stable template by which humans could understand their place in nature, as well as their value compared to other humans—kings being higher up the Chain than peasants, for instance. Then, in the seventeenth century, René Descartes did away with the many gradations of the *Scala* by cleaving the universe into just two modes of existence: *res cogitans* (mind stuff) and *res extensa* (matter stuff).

This sweeping simplification of nature's big picture brought with it many new problems. There was the metaphysical problem of how the two domains could ever interact—a question which has framed investigations of consciousness, for better or for worse, and largely for worse, ever since. There were disruptions, too, to the fine-grained ordering on which political and religious authority depended. If animals had elements of *res cogitans*, any signs of minds, what was to prevent them from aspiring to the spiritual realm, as humans could? And any attempt to intellectually investigate the soul—as Descartes seemingly advocated—was sure to annoy the powerful Catholic Church.

Descartes always played his cards carefully with the church, going as far as attempting to prove the existence of a benevolent God in his third and fifth *Meditations*. When it came to nonhuman animals, it is often claimed that he thought they lacked consciousness entirely. Although it is hard to be sure, this was probably not his view.* Descartes's primary claim about nonhuman animals was

---

* Descartes had a pet dog called Monsieur Grat (Mr. Scratch) to whom he was apparently extremely devoted. Then again, he also vivisected rabbits.

that they lacked souls and all the rational, spiritual, and conscious attributes that came along with having a soul. The historian Wallace Shugg summarized his views on the matter like this:

> The bodies of both man and beast . . . are merely machines that breathe, digest, perceive and move by means of the arrangement of parts. But only in man does reason direct bodily movements to meet all contingencies; only man gives evidence of his reason by using true speech. Without minds to direct their bodily movements or receive sensation, animals must be regarded as unthinking, unfeeling machines that move like clockwork.

In this view, the flesh-and-blood properties of living beings—their nature as organisms—are entirely and explicitly irrelevant to the presence of mind, consciousness, or soul (whatever that may be). Nonhuman animals are best thought of as *bêtes-machines*—in English, "beast machines." In the Cartesian picture, the division between mind and life is as sharp as that between *res cogitans* and *res extensa*.

By reinforcing the specialness of humans, Descartes was able to placate many of his would-be persecutors. But a dangerous door was now ajar. If nonhuman animals are beast machines, and if humans are a kind of animal too—humans, after all, certainly seemed to be made of the same sort of flesh, blood, cartilage, and bone—then surely the faculties of mind and reason should also be explicable in mechanistic, physiological terms?

The French philosopher Julien Offray de La Mettrie, writing in the middle of the eighteenth century, certainly saw things this way. He extended Descartes's beast machine argument to human beings, arguing that humans were machines too—*l'homme machine* (man machine)—and in doing so denying any special immaterial status for the soul while also questioning the existence of God. La Mettrie was not one to finesse his arguments for the benefit of religious authority, and so his life rapidly became a lot more complicated than Descartes's. In 1748 he was forced to flee his adopted home in the Netherlands to work for the Prussian King Frederick in Berlin, where three years later he died after consuming an excess of pâté.

While in the Cartesian view mind and life are independent, for La Mettrie they were deeply connected, in the sense that mind could be viewed as a property of life. Even today, discussions rumble on about whether life and mind are continuous or discontinuous, in terms of their underlying mechanisms and principles.

My sympathies in this debate lie with La Mettrie, but rather than speaking in general terms about "mind," my focus is on consciousness. This brings us to the heart of my "beast machine" theory of consciousness and self. *Our conscious experiences of the world around us, and of ourselves within it, happen with, through, and because of our living bodies.* Our animal constitution is not merely compatible with our conscious perceptions of self and world. My proposal is that we cannot understand the nature and origin of these conscious experiences, except in light of our nature as living creatures.

UNDERNEATH THE LAYERED EXPRESSIONS of selfhood involving memories of the past and plans for the future, before the explicit sense of personal identity, beneath the "I" and even prior to the emergence of a first-person perspective and experiences of body ownership, there are deeper layers of selfhood still to be found. These bedrock layers are intimately tied to the interior of the body, rather than to the body as an object in the world, and they range from emotions and moods—what psychologists call "affective" experiences—to a basal, formless, and ever-present sense of simply "being" an embodied, living organism.

We'll start our exploration of these depths with emotions and moods. These forms of conscious content are central to the experience of being an embodied self, and—like all perceptions—they, too, can be understood as Bayesian best guesses about the causes of sensory signals. The distinctive thing about affective experiences is that the relevant causes are to be found within the body, rather than out there in the world.

When we think about perception, we tend to think in terms of the different ways in which we sense the outside world—in particular the familiar modalities of sight, hearing, taste, touch, and smell. These world-oriented varieties of sensation and perception are collectively called *exteroception*. Perception of the body from within is known as *interoception*—it is the "sense of the internal physiological condition of the body."* Interoceptive sensory signals are typically transmitted from the body's internal organs—the viscera—to the central nervous system, conveying information about the state of those organs, as well as about the functioning of the body as a whole. Interoceptive signals report things like heartbeats, blood pressure levels, various low-level aspects of blood chemistry, degrees of gastric tension, how breathing is going, and so on. These signals travel

* Sitting in between exteroception and interoception is proprioception, which refers to the perception of body position and movement (see chapter 5). It's important not to confuse interoception with introspection, which refers to the internal examination of one's own mental states.

through a complex network of nerves and deep-lying brain regions in the brainstem and thalamus before arriving at parts of the cortex specialized for interoceptive processing—in particular the insular cortex.* The key property of interoceptive signals is that they reflect, in one way or another, how well physiological regulation of the body is going. In other words, how good a job the brain is doing of keeping its body alive.

Interoceptive signaling has long been connected with emotion and mood. Back in 1884, William James and—independently— Carl Lange argued that emotions were not the "eternal and sacred psychic entities" of the ancient philosophers, nor were they hard-wired into brain circuits by evolution, as Darwin had proposed not long before. Instead, they argued that emotions are perceptions of changes in bodily state. We don't cry because we are sad, we are sad because we perceive our bodily state in the condition of crying. The emotion of fear, in this view, is constituted by (interoceptive) perception of a whole gamut of bodily responses set off by the organism recognizing danger in its environment. For James, the perception of bodily changes as they occur *is* the emotion: "We feel sorry because we cry, angry because we strike, afraid because we tremble, and not that we cry, strike, or tremble, because we are sorry, angry, or fearful."

James's theory encountered strong resistance at the time, in part because it subverts the common, intuitive, how-things-seem notion that emotions cause bodily responses, rather than the other way around. It *seems as though* the feeling of fear—for example, when we happen upon a grizzly bear—is what causes our heart to race, our

---

* The insular cortex has its name because of its resemblance to an "island" within the larger cortical "sea."

adrenaline to pump, and our feet to flee. By now, though, we've learned to be skeptical about taking how things seem as a guide to how they actually are—so it would be unwise to dismiss the Jamesian view on this basis alone.

A more substantive concern was that bodily states may be insufficiently distinct from each other to support the full emotional range that we humans experience. While the specifics of this concern are still debated, a powerful response emerged in the 1960s with "appraisal theories" of emotion. On these theories, emotions are more than just readouts of changes in bodily state. They depend on a higher-level cognitive appraisal, or evaluation, of the context in which the physiological changes take place.

Appraisal theories solve the problem of emotional range because each specific emotion now no longer needs a dedicated bodily state. Two closely related emotions—for example, listlessness and ennui—might be based on the same bodily state. The distinct emotions would emerge from different cognitive interpretations of this shared bodily condition. Of course, it might equally be true—and I suspect that it probably is true—that every emotion does indeed have a distinct embodied signature, with the fine details of their distinguishing features just being very difficult to detect.

My favorite experimental investigation of appraisal theory comes from an inventive study by Donald Dutton and Arthur Aron in 1974, in which a female interviewer approached male passersby while they were crossing one of two bridges across the Capilano River in North Vancouver. One of the bridges was a 450-foot-long rickety suspension bridge with low handrails, precariously poised high above shallow rapids. The other was a shorter and sturdier affair made of heavy cedar, positioned farther upriver and only ten feet above the water. When the interviewer made contact with each bridge crosser, she invited him to fill out a questionnaire, and she also offered her phone number, explaining that she'd be happy to answer any further questions he might have.

The researchers wondered whether the men on the rickety bridge would misinterpret the physiological arousal caused by their precarious condition as sexual attraction, rather than as fear or anxiety. They reasoned that if this were so, these men would be more likely to call the interviewer after the event—maybe even ask her for a date.

This is exactly what happened. The female interviewer received more calls from men who had been crossing the rickety bridge than from those who had been crossing the sturdy bridge. Dutton and Aron called this a "misattribution of arousal": increased physiological arousal, induced by the rickety bridge, had been misinterpreted by higher-level cognitive systems as sexual chemistry. Supporting this appraisal theory interpretation (and assuming heterosexuality), when the questionnaire bearer was a man, there was no effect of type of bridge on the number of follow-up calls made.*

This study, conducted more than forty years ago, shows inevitable methodological weaknesses when compared to today's more rigorous, though still imperfect, standards. Not to mention the dubious ethics. But it still vividly illustrates the view that emotional

---

* In September 2020, I hiked across the notorious Sharp Edge ridge on the mountain Blencathra, in England's Lake District. Although no climbing gear is needed, traversing Sharp Edge is never easy. The ridge's crest is a jagged mess of slippery rock flanked by precipitous slopes, and accidents do happen. On this particular crossing, I noticed an upturned stone at the base of the ridge with the words "Marry me, Maria?" written in chalk. I couldn't help wonder whether whoever was responsible knew about Dutton and Aron's experiment, and was taking advantage.

experiences depend on how physiological changes are evaluated by higher-level cognitive processes.

One limitation of appraisal theories is that they assume a sharp distinction between what is taken as "cognitive" and what is not. Low-level "noncognitive" perceptual systems are assumed to "read out" the physiological condition of the body, while higher-level cognitive systems "evaluate" this condition through more abstract processes such as context-sensitive reasoning. For example, fear happens when a specific bodily state is first perceived, and then later evaluated as being "due to the presence of an approaching bear." However, unfortunately for appraisal theories, the brain does not separate neatly into "cognitive" and "noncognitive" domains.

Around 2010, as my research group at Sussex was getting off the ground, I started thinking about this problem. I'd been learning a lot about interoception from my colleague Hugo Critchley—one of the world's experts on the topic—and it occurred to me that a way to overcome the limitations of appraisal theories was to apply the principles of predictive perception, and to treat emotions and moods—and affective experiences in general—as distinctive kinds of controlled hallucination.

I called the idea "interoceptive inference". Just as the brain has no direct access to the causes of exteroceptive sensory signals like vision, which are out there in the world, it also lacks direct access to the causes of interoceptive sensory signals, which lie inside the body. *All* causes of sensory signals, wherever they are, are forever and always hidden behind a sensory veil. Interoception is therefore also best understood as a process of Bayesian best guessing, just like exteroceptive perception. In the same way that "redness" is the subjective aspect of brain-based predictions about how some surfaces reflect light, emotions and moods are the subjective aspects of predictions about the causes of interoceptive signals. They are internally driven forms of controlled hallucination.*

Just like visual predictions, interoceptive predictions operate at many scales of time and space, supporting fluid, context-sensitive, multilevel best guesses about the causes of interoceptive signals. In this way, interoceptive inference solves the problem of emotional range without needing any bright-line distinction between the noncognitive and the cognitive. Interoceptive inference is therefore more parsimonious than appraisal theory, because it involves just one process (Bayesian best guessing) rather than two (noncognitive perception and cognitive evaluation), and because of this, it also maps more comfortably onto the underlying brain anatomy.

Interoceptive inference is difficult to test experimentally, in part because it is harder to measure and manipulate interoceptive signals than is the case for exteroceptive modalities like vision. One promising approach explores the possibility that brain responses to heartbeats might be signatures of interoceptive prediction errors. The German neuroscientist Frederike Petzschner has recently shown that such responses, called "heartbeat evoked potentials," are modulated by paying attention, as predicted by interoceptive inference. More research along these lines is needed.

Another more indirect line of evidence comes from experiments on body ownership, like those we met in the previous chapter. In a 2013 study, led by Keisuke Suzuki, we found that people experienced greater ownership over a virtual reality "rubber hand" when

---

* In moving straight from appraisal theory to interoceptive inference I am skipping over a large body of intervening work. Antonio Damasio, in particular, has made seminal contributions showing how emotion and cognition are related, and how they both depend on the body. And Lisa Feldman Barrett independently came up with closely related ideas stressing the importance of interoceptive predictions.

it flashed in time with their heartbeat than when it flashed out of time—suggesting that body ownership depends on the integration of both exteroceptive and interoceptive signals. This "cardio-visual synchrony" method was also used by Jane Aspell and her colleagues in a "full body illusion" setup, in which people viewed a virtual silhouette of their body. They, too, found that people reported stronger identification with the silhouette when it flashed in time with the heartbeat. While these studies are suggestive of interoceptive inference, more research is needed here too, in part because these experiments didn't take account of individual differences in hypnotic suggestibility—a factor we've since learned is very important in body ownership experiments. They also depend on how aware a person is of their own heartbeat, a trait which has proven frustratingly difficult to measure.

From the perspective of the beast machine theory, the most important implication of interoceptive inference is that affective experiences are not merely shaped by interoceptive predictions but constituted by them. Emotions and moods, like all perceptions, come from the inside out, not the outside in. Whether it's fear, anxiety, joy, or regret—every emotional experience is rooted in top-down perceptual best guessing about the state of the body (and about the causes of this state). Recognizing this is the first key step toward understanding how experiences of being an embodied self are tied to our flesh-and-blood materiality.

To take the next step, we need to ask what these perceptions of the body "from within" are *for*. Perception of the outside world is obviously useful for guiding action, but why should our internal physiological condition be built into our conscious lives from the ground up? Answering this question takes us back in history once again, but this time only to the mid-twentieth century and to the neglected amalgam of computer science, artificial intelligence, engineering, and biology known as *cybernetics*.

IN THE 1950S, at the dawn of the computer age, the emerging disciplines of cybernetics and artificial intelligence (AI) were equally promising and in many ways inseparable. Cybernetics— from the Greek *kybernetes*, meaning "steersman" or "governor"— was described by one of its founders, the mathematician Norbert Wiener, as "the scientific study of control and communication in the animal and the machine." The emphasis of cybernetics was squarely on control, and its primary applications were in systems—such as guided missiles—that involved closed-loop feedback from output to input. One conspicuous feature of this approach was that such systems could appear to have "purposes" or "goals," like hitting a target.

This way of thinking about machines—as potentially having "purposes"—provided a new bridge from the nonliving to the living. Previously, the prevailing view had been that only biological systems could have goals, could behave according to an inner purpose.* Cybernetics suggested otherwise, emphasizing instead the close connections between machines and animals. Partly because of this, it split away from other approaches within AI which emphasized the off-line, disembodied, abstract reasoning that came to be exemplified by chess-playing computers. By most yardsticks these alternative approaches won the day, dominating the headlines

---

* I'm talking mainly about post-Enlightenment views. Earlier belief systems, such as animism, attributed purpose (and life, and spirit) far more liberally.

and the funding agencies, while cybernetics became increasingly relegated to the sidelines. Yet even in its relative obscurity, cybernetics delivered many valuable insights—the significance of which is only now becoming recognized.

One of these insights comes from a 1970 paper by William Ross Ashby and Roger Conant, which describes their so-called "Good Regulator Theorem." The concept is nicely encapsulated by the title of their paper: "Every good regulator of a system must be a model of that system."

Think about your central heating system, or—just as good— your air-conditioning system. Let's say this system is designed to keep the temperature inside your house at a steady 19°C (about 66°F). Most central heating systems work using simple feedback control: if the temperature is too low, switch on, otherwise switch off. Call this simple type of system "System A."

Imagine now a more advanced system, "System B." System B is able to *predict* how the temperature in the house would respond to the heating being on or off. These predictions are based on properties of the house—how big the rooms are, where the radiators are located, what the walls are made of—as well as on what the weather conditions are like outside. System B then adjusts the boiler output accordingly.

Thanks to these advanced abilities, System B is better at maintaining your house at a steady temperature than System A, especially if you have a complicated house or complicated weather. System B is better because it has a *model* of the house, which allows it to predict how the temperature inside the house will respond to the actions it can take. A top-end System B might even be able to anticipate upcoming temperature-related challenges—perhaps a cold day on the way—and alter the boiler output in advance, so as to guard against even a temporary drop in warmth. As Conant and Ashby said, *every good regulator of a system must be a model of that system.**

Let's take this example a step further. Imagine that System B has been fitted with imperfect "noisy" temperature sensors that only indirectly reflect the ambient temperature in the house. This means that the actual temperature cannot be directly "read off" from the sensors; instead, it has to be inferred on the basis of the sensory data and prior expectations. System B now has to have a model of (i) how its sensor readings relate to their hidden causes (the actual temperature in the house), and (ii) how these causes will respond to different actions, such as adjusting the boiler or radiator output.

We are now in a position to connect these ideas about regulation to what we know about predictive perception. System B works by inferring the ambient temperature from sensor readings, just as our brain makes best guesses about the causes of its sensory signals in order to infer states of the world (and body) and how they change over time. But the goal for System B is not to figure out "what's there"—in this case the ambient temperature. The goal is to *regulate* this inferred hidden cause, to take *action* so as to keep the temperature within a comfortable range, and ideally at a single fixed value. Perception, in this context, is not for figuring out what's there, it's for control and regulation.

Control-oriented perception—the sort of thing implemented by System B—is therefore a form of *active inference*, the process by which sensory prediction errors are minimized through making

---

* One might wonder whether there is a difference between "being" a model and "having" a model. I think that systems that possess explicit generative models capable of producing conditional or counterfactual predictions, like System B, can be said to "have" models. Regulators that are relatively fixed and inflexible, like a simple feedback thermostat, such as System A, may merely "be" a model.

actions rather than by updating predictions. As I explained in chapter 5, active inference depends both on generative models which are able to predict how the causes of sensory signals respond to different actions, and on modulating the balance between top-down predictions and bottom-up prediction errors, so that perceptual predictions can become self-fulfilling.

Active inference tells us that predictive perception can be geared either toward inferring features of the world (or the body) or toward regulating these features—it can be about finding out things or about controlling things. What cybernetics brings to the table is the idea that, for some systems, control comes first. From the perspective of the Good Regulator Theorem, the entire apparatus of predictive perception and active inference emerges from a fundamental requirement about what it takes to adequately regulate a system.

To answer the question of what perceptions of emotion and mood are for, we need one more concept from cybernetics—that of an *essential variable*. Also introduced by Ross Ashby, essential variables are physiological quantities, such as body temperature, sugar levels, oxygen levels, and the like, that must be kept within certain rather strict limits in order for an organism to remain alive. By analogy, a desired room temperature would be the "essential variable" for a central heating system.

Putting these pieces together, emotions and moods can now be understood as *control-oriented perceptions which regulate the body's essential variables*. This is what they are *for*. The experience of fear I feel as a bear approaches is a control-oriented perception of my body—more specifically "my body in the presence of an approaching bear"—that sets off a series of actions that are best predicted to keep my essential variables where they need to be. Importantly, these actions can be both external movements of the body—like running—and internal "intero-actions" such as raising the heart rate or dilating blood vessels.

This perspective on emotions and moods ties them even more closely to our flesh-and-blood nature. These forms of self-perception are not merely about registering the state of the body, whether from the outside or from the inside. They are intimately and causally bound up with how well we are doing, and how well we are likely to do in the future, at the business of staying alive.

Crucially, in making this distinction, we also find the reason why emotions and moods have their characteristic phenomenology. Experiences of fear, jealousy, joy, and pride are very different, but they are more similar to each other than any one of them is to a visual experience, or to an auditory experience. Why is this? The nature of a perceptual experience depends not only on the target of the corresponding prediction—perhaps a coffee cup on the table, or a racing heart—but also on the type of prediction being made. Predictions geared toward finding out things will have a very different phenomenology from those geared toward controlling things.

When I look at a coffee cup on my desk, there is the strong perceptual impression of a three-dimensional object that exists independently of me. This is the phenomenology of "objecthood," which I introduced in chapter 6. There, I proposed that objecthood arises in visual experience when the brain makes conditional predictions about how visual signals would change, given this-or-that action—like rotating a cup to reveal its back. Perceptual predictions in this case are geared toward finding out what's there, and the relevant actions are those, like rotations, that are predicted to reveal more about the hidden causes of the sensory signals.

Now consider a more active example: catching a cricket ball. You might think the best way to do this would be to figure out where the ball is going to land and run there as fast as possible. But "figuring out what's there" is not, in fact, a good strategy, and is not what expert fielders do. Instead, you should keep moving so that the ball always "looks the same" in a particular way—specifically, so that the angle of elevation of your gaze to the ball increases, but at a steadily decreasing rate. It turns out that if you follow this strategy—psychologists call it "optic acceleration cancellation"—you are guaranteed to intercept the ball.*

This example brings control back into the picture. Your actions—and your brain's predictions about their sensory consequences—are not about finding out where the ball *is*. They are geared toward controlling how the ball perceptually *appears*. Accordingly, your perceptual experience will not reveal the precise location of the ball in the air, but something like its "catchability" as you run toward it. Perception in this situation is a *controlling* hallucination just as much as it is a *controlled* hallucination.

This idea has substantial historical pedigree. In the 1970s, the psychologist James Gibson argued that we often perceive the world in terms of what he called "affordances." An affordance, for Gibson, is an *opportunity for action*—a door for opening, a ball for catching—rather than an action-independent representation of the "way things are." Another theory, also from the 1970s but less well-known than Gibson's, puts even more emphasis on control. According to William Powers's "perceptual control theory," we don't perceive things in order to then behave in a particular way. Instead, as in the example of catching a cricket ball, we behave so that we end up perceiving things in a particular way. While these early theories were conceptually on track, and are in line with the "action first" view of

---

* If you take this advice literally, the ball will end up hitting you right between the eyes.

the brain that I introduced in chapter 5, they lacked the concrete predictive mechanisms provided by the controlled hallucination—or controlling hallucination—view of perception. They also focused on perception of the outside world, rather than on the interior of the body.

Anxiety doesn't have a back, sadness doesn't have sides, and happiness is not rectangular. The perceptions of the body "from within" on which affective experiences are built do not deliver experiences of the shape and location of my various internal organs—my spleen here, my kidneys over there. There is no phenomenology of objecthood, as when looking at a coffee cup on the table, nor is there anything like movement in a spatial frame, as when catching a cricket ball.

The control-oriented perceptions that underpin emotions and moods are all about predicting the consequences of actions for keeping the body's essential variables where they belong. This is why, instead of experiencing emotions as objects, we experience how well or badly our overall situation is going, and is likely to go. Whether I'm sitting by my mother's hospital bed, or fixing to escape from a bear, the form and quality of my emotional experiences are the way they are—desolate, hopeful, panicky, calm—because of the conditional predictions my brain is making about how different actions might impact my current and future physiological condition.

AT THE VERY DEEPEST LAYERS of the self, beneath even emotions and moods, there lies a cognitively subterranean, inchoate, difficult-to-describe experience of simply *being a living organism*. Here, experiences of selfhood emerge in the unstructured feeling of just "being." This is where we reach the core of the beast machine theory: the proposal that conscious experiences of the world around us, and of ourselves within it, happen *with*, *through*, and *because of* our living bodies. It is at this point that all of the ideas I've been putting forward about perception and self fall into place. So let's take things step by step, from the beginning.

The primary goal for any organism is to continue staying alive. This is true almost by definition—an imperative endowed by evolution. All living organisms strive to maintain their physiological integrity in the face of danger and opportunity. This is why brains exist. Evolution's reason for providing organisms with brains is not so they can write poetry, do crossword puzzles, or pursue neuroscience. Evolutionarily speaking, brains are not "for" rational thinking, linguistic communication, or even for perceiving the world. The most fundamental reason any organism has a brain—or any kind of nervous system—is to *help it stay alive*, through making sure that its physiological essential variables remain within the tight ranges compatible with its continued survival.

These essential variables, whose effective regulation determines the life-status and future prospects of an organism, are the causes of interoceptive signals. Like all physical properties, these causes remain hidden behind a sensory veil. Just as with the outside world, the brain has no direct access to physiological states of the body, and so these states have to be inferred through Bayesian best guessing.

As with all predictive perception, this best guessing is achieved through a brain-based process of prediction error minimization. In the context of interoception, this is called interoceptive inference. Just as with vision and with hearing—just as with *all* perceptual modalities—interoceptive perception is a kind of controlled hallucination.

Whereas perceptual inference about the world is often geared toward finding out things, interoceptive inference is primarily about controlling things—it is about physiological regulation. Interoceptive inference exemplifies active inference, in that prediction errors are minimized by acting to fulfill top-down predictions, rather than by updating the predictions themselves (though this happens too). These regulatory actions can be external, like reaching for food, or internal, like gastric reflexes or transient alterations in blood pressure.

This kind of predictive control can support *anticipatory* responses, through predictions about future bodily states and their dependence on this-or-that action. This kind of anticipatory control can be critical for survival. For example, it may turn out very badly to wait for something like blood acidity to go out of bounds before mustering an appropriate response. Again, the relevant actions can be external, internal, or both. Running away from a bear before being eaten is an example of external anticipatory regulation. The transient increase in blood pressure needed to run away effectively, or even to stand up from your desk after you've been working for a while, is an internal anticipatory response.

There is a useful term in physiology to describe this process: "allostasis." Allostasis means the process of achieving stability through change, as compared to the more familiar term "homeostasis," which simply means a tendency toward a state of equilibrium. We can think of interoceptive inference as being about the allostatic regulation of the physiological condition of the body.

Just as predictions about visual sensory signals underpin visual experiences, interoceptive predictions—whether about the future, or about the here and now—underpin emotions and moods. These affective experiences have their characteristic phenomenology because of the control-oriented and body-related nature of the perceptual predictions that they depend on. They are controlling hallucinations just as much as they are controlled hallucinations.

Despite being firmly rooted in physiological regulation, emotions and moods are still mostly experienced at least in part as relating to things and situations beyond the self, outside the body. When I feel fear, I am usually afraid of some *thing*. But the very deepest levels of experienced selfhood—the inchoate feeling of "just being"—seem to lack these external referents altogether. This, for me, is the true ground-state of conscious selfhood: a formless, shapeless, control-oriented perceptual prediction about the present and future physiological condition of the body itself. This is where *being you* begins, and it is here that we find the most profound connections between life and mind, between our beast machine nature and our conscious self.

The final, and crucial, step in the beast machine theory is to recognize that from this starting point, everything else follows. We are not the beast machines of Descartes, for whom life was irrelevant to mind. It is exactly the opposite. *All* of our perceptions and experiences, whether of the self or of the world, are inside-out controlled and controlling hallucinations that are rooted in the flesh-and-blood predictive machinery that evolved, develops, and operates from moment to moment always in light of a fundamental biological drive to stay alive.

We are conscious beast machines, through and through.

AT THE END OF the previous chapter, I noted that while perceptions of the world come and go, experiences of selfhood seem to be stable and continuous over many different timescales. We can now see that this subjective stability emerges naturally from the beast machine theory.

To effectively regulate the body's physiological condition, priors on interoceptive signals need to have high precision, so that they tend to become self-fulfilling. This key aspect of active inference ensures that interoceptive best guesses will be drawn toward these priors—to the desired (predicted) regions of physiological viability. For example, my body temperature is predicted to be rather constant over time, which—according to active inference—is why it actually turns out this way. The experience of the bodily self as being relatively unchanging therefore stems directly from the need to have precise priors—strong predictions—about stable bodily states, for the purposes of physiological regulation. To put it another way: for as long as we live, the brain will never update its prior belief of expecting to be alive.*

What's more, given that "change" is itself an aspect of perceptual inference, the brain may attenuate prior expectations related to perceiving changes in the condition of the body, in order to further ensure that physiological essential variables stay where they ought to be. This implies a form of "self-change-blindness"—a notion also introduced in the previous chapter. In this view, we might not perceive our physiological condition as changing even when it does, in fact, change.

Putting these ideas together, we perceive ourselves as stable over time in part because of a self-fulfilling prior expectation that our physiological condition is restricted to a particular range, and in

---

* Another way to think of this is that interoceptive sensory signals will be systematically "disattended to" in order to allow intero-actions to regulate essential variables, in just the same way that proprioceptive sensory signals are attenuated during external actions—as discussed in chapter 5.

part because of a self-fulfilling prior expectation that this condition does not change. In other words, effective physiological regulation may depend on systematically *mis*perceiving the body's internal state as being more stable than it really is, and as changing less than it really does.

Intriguingly, this proposal may generalize to other, higher levels of selfhood beyond the ground-state of continued physiological integrity. We will be better able to maintain our physiological and psychological identity, at every level of selfhood, if we do not (expect to) perceive ourselves as continually changing. Across every aspect of being a self, we perceive ourselves as stable over time because we perceive ourselves in order to control ourselves, not in order to know ourselves.

Complementing this subjective stability, most of us most of the time also perceive ourselves as being "real." This may seem obvious, but remember from chapter 6 that the experience of things in the world as "really existing" is not evidence of direct perceptual access to an objective reality, but a phenomenological property that needs to be explained. There, I proposed that to be useful for the perceiving organism, our perceptual best guesses need to be experienced as *really existing out there in the world*, rather than as the brain-based constructions that in truth they are.

The same reasoning holds for the self too. Just as it seems as though the chair in the corner *really is* red, and that a minute *really has* passed since I started writing this sentence, the predictive machinery of perception when directed inwardly makes it seem as though there *really is* a stable essence of "me" at the center of everything.

And in the same way that our perceptions of the world can sometimes lack the phenomenology of being real, the self, too, can lose its reality. The experienced reality (and subjective stability) of the self may wax and wane during illness, and it can be severely attenuated or even abolished in the psychiatric condition of depersonalization. The most extreme examples of self-unreality happen in a rare delusion first described in 1880 by the French neurologist Jules Cotard. The embodied self is so far gone in the Cotard delusion that sufferers believe they do not exist, or that they are already dead. Of course, the *experience* that the self is unreal does not mean that any essence-of-self has upped and left. It just means that the control-oriented perceptions associated with the deepest layers of bodily regulation have gone significantly awry.

IN PUTTING FORWARD this beast machine theory, I am not claiming to have demonstrated that life is necessary for consciousness; that there is something special about flesh, blood, and guts—or biological neurons—which means that only creatures built from these materials can have conscious experiences. This may be true, or it may not. Nothing I've said, at least so far, makes a strong case either way, nor is it intended to. What I *am* claiming is that in order to understand why our conscious experiences are the way they are, what experiences of self are like, and how they relate to experiences of the world, we will do well to appreciate the deep roots of all perception in the physiology of the living.

Thinking about the material basis of consciousness brings us back, once again, to the hard problem. The beast machine theory accelerates the dissolution of this apparent mystery. By extending the controlled hallucination view to the very deepest layers of selfhood, by revealing the experience of *the-self-as-really-existing* as one more aspect of perceptual inference, the intuitions on which the hard problem implicitly rest are eroded even further. In particular, the hard-problem-friendly intuition that the conscious self is somehow apart from the rest of nature—a really-existing immaterial inner observer looking out onto a material external world—turns out to be just one more confusion between how things *seem* and how they *are*.

Centuries ago, when Descartes and La Mettrie were forming their views about the relations between life and mind, it was not the hard problem that was at issue, but the existence—or nonexistence—of the "soul." And—perhaps surprisingly—there are echoes of the soul to be found in the beast machine story too. This soul is not an immaterial quiddity, nor a spiritual distillation of rationality. The beast machine view of selfhood, with its intimate ties to the body, to the persistent rhythms of the living, returns us to a place liberated from conceits of a computational mind, before Cartesian divisions of mind and matter, reason and non-reason. What we might call the "soul" in this view is the perceptual expression of a deep continuity between mind and life. It is the experience we have when we encounter the deepest levels of embodied selfhood—these inchoate feelings of "just being"—as *really existing*. It seems right to call this an echo of the soul because it revives even more ancient conceptions of this eternal notion, conceptions—such as the Ātman in Hinduism—which contemplated our innermost essence more as breath than as thought.

We are not cognitive computers, we are feeling machines.

10

A FISH IN WATER

IN SEPTEMBER OF 2007, I was on my way from Brighton to Barcelona to give a talk at a summer school on "brain, cognition, and technology." Although I was happy to be traveling to such a beautiful city, duties at home meant I'd arrive too late to attend a three-hour master class by the eminent British neuroscientist Karl Friston on his "free energy principle" and its application to neuroscience. (Friston made an appearance in chapter 5, when we met his concept of active inference.) I'd been eager to hear Friston's seminar because his ideas seemed to capture, in a mathematically profound albeit complicated way, some of my own embryonic thoughts about predictive perception and the self.

Resigned to missing his talk, I thought I'd at least be able to find out what had happened when I got there. But when I turned up at the rooftop bar later that evening, I was met by a sea of bemused faces. Karl himself had hopped on a plane back to London immediately after his lecture, leaving bafflement in his wake. It turned out that after three hours of detailed mathematics and neuroanatomy, most people were even more confused than they'd been to begin with.

Part of the problem seemed to be the sheer scale of what was being proposed. The first thing that strikes you about the free energy principle is that it's a really big idea. It brings together concepts, insights, and methods from biology, physics, statistics, neuroscience, engineering, machine learning, and elsewhere besides. And its application is by no means limited to the brain. For Friston, the free energy principle explains *all* features of living systems, from the self-organization of a single bacterium, to the fine details of brains and nervous systems, to the overall shape and body plan of animals, reaching even as far as the broad strokes of evolution itself. It's as close to a "theory of everything" in biology as has yet been proposed. It is no wonder that people—me included—were bewildered.

Fast-forward ten years. In 2017, my colleagues Chris Buckley, Simon McGregor, Chang-Sub Kim, and I finally published our own review of "the free energy principle in neuroscience," in the *Journal of Mathematical Psychology*. Getting there had taken us about nine years longer than anticipated, but I'm glad we persevered.

At least I *think* I'm glad, since even after all those hard yards, there remains something strangely inscrutable about it all. Across the internet, blog posts regularly report on struggles to comprehend Friston's ideas. There's Scott Alexander's "God help us, let's try to understand Friston on free energy." There's even a parody Twitter account, @FarlKriston, which posts gnomic statements like "I am, whatever I think I am. If I wasn't, why would I think I am?"

But the free energy principle is worth it, because accompanying its apparent inscrutability, there is an elegance and simplicity that points to a deep unity between life and mind, and which in doing so fills out the beast machine theory of consciousness in several important ways.

And as we'll see, when boiled down far enough, the free energy principle—the FEP for short—is not so hard to understand after all.

LET'S PUT ASIDE THE MYSTERIOUS "free energy" for a moment, and begin with a simple statement about what it means for an organism, indeed for anything, to exist.

What it means for something to exist is that there must be a difference—a boundary—between that thing and everything else. If there were no boundaries, there would be no things—there would be nothing.

This boundary must also persist over time, because things that exist maintain their identity over time. If you add a drop of ink to a glass of water, it will rapidly disperse, coloring the water and losing its identity. If, instead, you add a drop of oil, although the oil will spread out over the surface, it will remain recognizably separate from the water. The oil drop continues to exist because it has not dispersed itself evenly throughout the water. After a while, though, it too will lose its identity, just as rocks eventually erode into dust. Things like oil drops and rocks undoubtedly exist, because they have an identity that persists for some period of time—a long time, for rocks. But neither oil drops nor rocks *actively* maintain their boundaries, they just get dispersed slowly enough for us to notice them as existing while this happens.

Living systems are different. Unlike the examples above, living systems actively maintain their boundaries over time—through moving, or sometimes even just through growing. They actively contribute to preserving themselves as distinct from their environment, and this is a key feature of what makes them living. The starting point for the FEP is that living systems, simply by virtue of existing, must actively resist the dispersion of their internal states. By the time you end up as a puddle of undifferentiated mush on the floor, you are no longer alive.*

Thinking about life this way brings us back to the concept of *entropy*. In chapter 3, I introduced entropy as a measure of disorder, diversity, or uncertainty. The more disordered a system's states are— like an ink drop dispersed messily throughout the water—the higher the entropy. For you, or me, or even a bacterium, our internal states are less disordered when we are alive than when we decompose into mush. Being alive means being in a condition of *low entropy*.

Here's the problem. In physics, the second law of thermody-namics tells us that the entropy of any isolated physical system in-creases over time. All such systems tend toward disorder, toward a dispersion of their constituent states over time. The second law tells us that instances of organized matter, like living systems, are intrin-sically improbable and unstable, and that—in the long run—we're all doomed. But somehow, unlike rocks or ink drops, living systems temporarily fend off the second law, persisting in a precarious con-dition of improbability. They exist out of equilibrium with their en-vironment, and this is what it means to "exist" in the first place.

According to the FEP, for a living system to resist the pull of the second law, it *must occupy states which it expects to be in*. Being a Good Bayesian, I'm using "expect" in a statistical sense, not in a psychological sense. It is a very simple, almost trivial idea. A fish in water is in a state it statistically expects to be in, because most fish are indeed in water most of the time. It is statistically unexpected to find a fish out of water, unless that fish is beginning to turn to

---

* There are some fascinating edge cases which are typically not considered as living, but which nonetheless seem to actively maintain their identities—consider a tornado, or a whirlpool.

mush. My body temperature being roughly 37°C is also a statistically expected state, compatible with my continued survival, with my not dissolving into mush.

For any living system, the condition of "being alive" means proactively seeking out a particular set of states that are visited repeatedly over time, whether these are body temperatures, heart rates (the physiological "essential variables" that we met in the previous chapter), or the organization of protein complexes and energy flows in a single-celled bacterium. These are the statistically expected, low-entropy states that ensure the system stays alive—the kind of states that are expected, given the creature in question.*

Importantly, living systems are not closed, isolated systems. Living systems are in continual open interaction with their environments, harvesting resources, nutrients, and information. It is by taking advantage of this openness that living systems are able to engage in the energy-thirsty activity of seeking out statistically expected states, minimizing entropy, and warding off the second law.

From the perspective of an organism, the entropy that matters is the entropy of its *sensory* states—those states that bring it into contact with its environment. Imagine a very simple living system, like a single bacterium. This bacterium requires a particular nutrient to survive, and it can sense the concentration of this nutrient in its immediate environment. By *expecting* to sense high nutrient concentrations, and by actively seeking out such expected sensory signals through its movements, this simple organism will maintain

---

* How can statistically expected states also be improbable? This is possible when a system inhabits only a restricted repertoire or subset of states—a so-called "attracting set"—out of a large number of possible states. The attracting set is statistically expected because that is where the system is usually found, but it is also improbable because there are many more states outside the set than inside it. There are many more ways of being mush than there are of being alive.

itself in the set of states that define it as being alive. In other words, sensing high nutrient concentrations is a statistically expected state for the bacterium, which it proactively seeks to keep visiting.

According to the FEP, this applies across the board. Ultimately, *all* organisms—not just bacteria—stay alive by minimizing their sensory entropy over time, thereby helping to ensure that they remain in the statistically expected states compatible with survival.

Here's where we get to the core of the FEP, which addresses the question of how, in practice, living systems manage to minimize their sensory entropy. Normally, to minimize a quantity, a system has to be able to measure it. The problem here is that sensory entropy cannot be directly detected or measured. A system cannot "know" whether its own sensations are surprising, simply on the basis of the sensations themselves. (Here's an analogy: Is the number 6 surprising? It's impossible to say, without knowing the context.) This is why sensory entropy is very different from things like levels of light, or concentrations of nearby nutrients, which *can* be directly detected by an organism through its senses, and used to guide behavior.

This is where *free energy* finally enters the story. Don't worry about the name, which has its origins in nineteenth-century theories of thermodynamics.* For our purposes, we can think of free energy as a quantity which approximates sensory entropy. Crucially, it is also a quantity that can be measured by an organism—and therefore it is something that the organism can minimize.

Following the FEP, we can now say that organisms maintain

---

* In thermodynamics, free energy is the amount of energy available for doing work at a constant temperature. It is "free" in the sense of being "available." The kind of free energy in the FEP is called "variational free energy"—a term which comes from machine learning and information theory, but which is closely related to its thermodynamic equivalent.

themselves in the low-entropy states that ensure their continued existence by actively minimizing this measurable quantity called "free energy." But what is free energy from the perspective of the organism? It turns out, after some mathematical juggling, that free energy is basically the same thing as sensory *prediction error*. When an organism is minimizing sensory prediction error, as in schemes like predictive processing and active inference, it is also minimizing this theoretically more profound quantity of free energy.

One implication of this connection is that the FEP licenses the idea from the previous chapter that living systems have—or are—models of their environment. (More specifically, models of the causes of their sensory signals.) This is because in predictive processing, as we saw in chapter 5, models are needed to supply the predictions that in turn define prediction errors. According to the FEP, it is in virtue of having or being a model that a system can judge whether its sensations are (statistically) surprising. (If you believe that the number 6 you see comes from the roll of a die, you can judge exactly how surprising it is.)

These deep connections between the FEP and predictive processing make appealing sense. Intuitively, by minimizing prediction error through active inference, living systems will naturally come to be in states they expect—or predict—themselves to be in. Seen this way, the ideas of predictive perception and controlled (or controlling) hallucination follow seamlessly from Friston's ambitious attempt to explain the whole of biology.

Putting all this together, the picture that emerges is of a living system actively modeling its world and its body, so that the set of states that define it as a living system keep being revisited, over and over again—from the beating of my heart every second to commiserating my birthday every year. Paraphrasing Friston, the view from the FEP is of organisms gathering and modeling sensory information so as to maximize the sensory evidence for their own existence. Or, as I like to say, "I predict myself, therefore I am."

It's worth noting that minimizing free energy—sensory prediction error—does *not* mean that a living system can get away with retreating into a dark and silent room and staying there, staring at the wall. You might think this would be an ideal strategy, since sensory inputs from the external environment will become highly predictable. But it is far from ideal. Over time, sensory inputs signaling other things, like levels of blood sugar and so on, will start to deviate from their expected values: you're going to get hungry if you stay in the dark room too long. Sensory entropy will start to grow, and nonexistence will loom. Complex systems like living organisms need to allow some things to change in order for other things to stay the same. We have to move to get out of bed and make breakfast, and our blood pressure has to rise while doing so, so that we don't faint. This matches the anticipatory form of predictive control—allostasis—that I mentioned in the previous chapter. Minimizing sensory prediction error in the long run means getting out of the dark room, or at least switching on the lights.

Another common worry about the FEP is that it is not falsifiable, in the sense that it cannot be proven wrong by experimental data. This is true, but it is neither unique to the FEP nor particularly problematic. The best way to think of the FEP is as a piece of mathematical philosophy rather than a specific theory that can be evaluated by hypothesis testing. As my colleague Jakob Hohwy puts it, the FEP addresses the question "What are the conditions for the possibility of existence?" in the same first-principles way that Immanuel Kant raised the question "What are the conditions for the possibility of perception?" The role of the FEP can be understood as motivating and facilitating the interpretation of other, more specific theories—theories which *are* amenable to refutation by experiment. The theory of predictive processing, for example, can be falsified if it turns out that the brain doesn't use sensory prediction errors in the process of perceiving. In the end, the FEP will be judged on how useful it is, not on whether it is empirically true or false itself.*

Let's summarize the main steps of the FEP. In order for organisms to stay alive they need to behave so as to maintain themselves in the (low entropy) states they "expect" to be in. A fish swimming above a coral reef, searching for food, is proactively seeking expected sensory states compatible with its continued survival. In general, living systems do this by minimizing a measurable approximation of the entropy of these states, which is free energy. Minimizing free energy requires the organism to have, or to be, a model of its environment (which includes the body). Free-energy-minimizing organisms then use these models to reduce the difference between predicted and actual sensory signals, by updating predictions and by performing actions. Indeed, given plausible mathematical assumptions, free energy turns out to be exactly the same thing as prediction error. Altogether, this means that the entirety of predictive processing and controlled hallucinations, of active inference and control-oriented perception—and of the beast machine theory too—can be understood through the lens of the FEP as flowing from a fundamental constraint on what it means to be alive, on what it means to exist.

---

* Another example of a principle like the FEP is Hamilton's "principle of stationary action" in physics, which can be used to derive (testable) equations of motion and even general relativity.

*   *   *

IN CASE YOU'VE FOUND this rapid take on the FEP a bit disorienting, let me reassure you that it is not necessary to comprehend or accept the FEP in order to follow the story of controlled hallucinations and beast machines as I've laid it out in previous chapters.* The theory that we experience the world, and the self, through mechanisms of predictive perception that are rooted in a "drive to stay alive" stands up all by itself. However, the FEP is worth the journey because it enhances the beast machine theory in at least three significant ways.

First, the FEP grounds the beast machine theory in the bedrock of physics, and in particular within a physics relevant to what it means to be alive. The beast machine's "drive to stay alive" resurfaces in the FEP as an even more fundamental imperative to remain in statistically expected states, to withstand the insistent pull of the second law of thermodynamics. When a theory can be generalized and grounded this way, it becomes more compelling, more integrative, and more powerful.

Second, the FEP firms up the beast machine theory by retelling it in reverse. Over the previous chapters, we began with the challenge of inferring what the outside world is like from within the vault of a bony skull, and then following the thread of ideas inward to the body—dealing first with experiences of selfhood as perceptual best guesses, and finally identifying the most deep-seated of these experiences with control-oriented perception of the body

---

* The concepts and mathematics behind the FEP are not simple, even for those with expertise in the area. The opening lines of a textbook on statistical mechanics warn us: "Ludwig Boltzmann, who spent much of his life studying statistical mechanics, died in 1906, by his own hand. Paul Ehrenfest, carrying on the work, died similarly in 1933. Now it is our turn to study statistical mechanics."

itself. With the FEP, it's the other way around. We start with the simple statement that "things exist" and proceed outward from there to the body and to the world. Arriving at the same place from two very different starting points strengthens the intuition that the underlying story is coherent, and makes clear otherwise obscure parallels between concepts (for example, free energy and prediction error).

The third benefit of the FEP lies in the rich mathematical toolbox that it brings to the table. This toolbox offers many new opportunities to further develop the ideas I've presented in previous chapters. Here's one example. When we unpack the mathematics of the FEP in more detail, we discover that what I really need to do, in order to stay alive, is to minimize free energy *in the future*—not just in the here and now. And it turns out that minimizing this long-term prediction error means I need to seek out new sensations *now* that reduce my uncertainty about what would happen *next*, if I did such-and-such. I become a curious, sensation-seeking agent—not someone content to self-isolate in a dark room. The mathematics of the FEP helps quantify this fine balance between exploration and exploitation, and this in turn has implications for what we perceive, since what we perceive is always and everywhere built from the predictions the brain is making. Insights like this will enable us to do better experiments, build sturdier explanatory bridges to carry the weight of these experiments, and, bit by bit, bridge by bridge, bring us ever closer to a satisfying explanation of how mechanisms give rise to minds.

At the same time, the FEP, despite being touted as a "theory of everything," is not a theory of consciousness. The FEP bears the same relationship to consciousness as do predictive, Bayesian theories of the brain: they are theories *for* consciousness science, in a real problem sense, and not *of* consciousness, in the hard problem sense. The FEP brings new insights and tools to the challenge of explaining phenomenology in terms of mechanism. And in return, notions of controlled hallucinations and beast machines endow the austere mathematics of the FEP with a newfound relevance for consciousness—and what good is a theory of everything unless it has something to say about that?

MANY YEARS AFTER my first unsettling encounter with the FEP, I spent a few days with Karl Friston—and about twenty other neuroscientists, philosophers, and physicists—at a small gathering on the Greek island of Aegina, an hour's ferry ride from Athens. It was another September, this time in 2018, not long after my mother's journey through delirium. Just as with the trip to Barcelona more than a decade earlier, I'd been looking forward to some late summer sun along with the science. The plan was to discuss the FEP with a focus on its relationship to the integrated information theory of consciousness (IIT)—the equally ambitious theory we explored in chapter 3. But instead of warm sunshine and blue skies, we were greeted by a major storm, a rare "medicane," which tossed tables and chairs into the sea and whipped the normally placid Mediterranean into a white-water fury.

As we sat in the conference annex, doors slamming in the gusts and branches lashing the windows, it struck me how extraordinary it was to be working on consciousness at a time when we had two highly ambitious and mathematically detailed theories which did not seem to speak to each other at all. On the face of it, this lack of interaction might have been disheartening, but I found it a fascinating situation to be in.

The storm continued to batter us throughout the day. Some ideas were floated, but I had the feeling we were mostly casting around in semidarkness. The FEP and IIT are both grand theories, but they are grand in different ways. The FEP starts from the simple statement that "things exist" and derives from this the whole of neuroscience and biology, but not consciousness. IIT starts from the simple statement "consciousness exists" and from there launches a direct assault on the hard problem. It's not surprising that they often talk past each other.

Two years later, as I put the finishing touches on this book, the two theories still live in different worlds. But now there are at least some tentative attempts underway to compare their experimental predictions. The planning discussions for these experiments—which I'm fortunate to be part of—have been by turns illuminating and frustrating, largely because of the dramatically different starting points and explanatory goals that each theory brings to the table. How these experiments will turn out remains to be seen. My intuition is that we will learn many useful things, but that neither the FEP nor the IIT will be explicitly ruled out as a theory of, or for, consciousness.

My own ideas about controlled hallucinations and beast machines chart a middle course. They share with the FEP a deep theoretical grounding in the nature of the self, and they leverage the powerful mathematical and conceptual machinery of the predictive brain. They share with IIT a clear focus on the subjective, phenomenological properties of consciousness—though with the real problem, not the hard problem, in the crosshairs. Rather than pitting the FEP against IIT, my hope is that the beast machine theory of consciousness and self provides a way to bring them together, weaving insights from both into a satisfying picture of why we are what we are.

Back on Aegina, the meeting ended, as most do, without any great fanfare. By the time we caught the ferry back to Athens, the storm had abated and the sea was quiet. It had been a difficult decision to make this trip. I'd missed some personally significant events back in Brighton. But in the end I'd decided, and standing on the deck in the sunshine, I was now at peace with that decision. I started thinking about how I'd made the decision, why it's always so hard, and before long I was thinking about how anyone makes any decision, and about what it means for us to feel in control of our choices, of our behavior, at all.

Once you start thinking about free will, there's really no stopping.

11

DEGREES OF FREEDOM

She bent her finger and then straightened it. The mystery was in
the instant before it moved, the dividing moment between not
moving and moving, when her intention took effect. It was like a
wave breaking. If she could only find herself at the crest, she
thought, she might find the secret of herself, that part of her that
was really in charge. She brought her forefinger closer to her face
and stared at it, urging it to move. It remained still because she
was pretending . . . And when she did crook it finally, the action
seemed to start in the finger itself, not in some part of her mind.

IAN McEWAN, *ATONEMENT*

WHAT IS THE ASPECT of *being you* that you cling to most tightly?
For many, it's the feeling of being in control of your actions, of being
the author of your thoughts. It's the compelling but complex notion
that we act according to our own free will.

Ian McEwan finds this complexity even in the simple flexing of
a finger. Thirteen-year-old Briony Tallis feels that her conscious intentions, for example to bend a finger, cause physical actions—the
actual bending of the finger. The line of apparent causation goes
straight from conscious intention to physical action. And she feels

that in this process lies the very essence of selfhood, of what it is to be her. But when Briony delves deeper into these feelings, things are not so simple. Where did the movement start? In the mind, or in the finger? Did the intention—or her "self"—cause the action, or was the experience of intention a result of perceiving the finger begin to move?

Pondering these questions, Briony Tallis has plenty of company. Few topics in philosophy and neuroscience have been as consistently inflammatory as free will. What it is, whether it exists, how it happens, whether it matters—consensus on these issues has remained elusive to say the least. There is not even clarity about the *experience* of free will—whether it is a singular experience or a class of related experiences, whether it differs among people, and so on. But amid all this confusion, there is one stable intuition. When we exercise free will, there is—in the words of the philosopher Galen Strawson—a feeling of "radical, absolute, buck-stopping *up-to-me-ness* in choice and action." A feeling that the self is playing a causal role in action in a way that isn't the case for a merely reflexive response, such as when you withdraw your hand from the sting of a nettle. This is why experiences of free will go naturally along with voluntary actions—whether flexing your finger, deciding to make a cup of tea, or embarking on a new career.

When I experience "freely willing" an action, I am in some sense experiencing *my self* as the cause of that action. Perhaps more than any other kind of experience, experiences of volition make us feel that there is an immaterial conscious "self" pulling strings in the material world. This is how things seem.

But experiences of volition do not reveal the existence of an immaterial self with causal power over physical events. Instead, I believe that they are distinctive forms of self-related perception.

More precisely, that they are self-related perceptions associated with voluntary actions. Like all perceptions—whether self-related or world-related—experiences of volition are constructed according to the principles of Bayesian best guessing, and they play important—likely essential—roles in guiding what we do.

Let's first be clear about what free will is *not*. Free will is not an intervention in the flow of physical events in the universe, more specifically in the brain, making things happen that wouldn't otherwise happen. This "spooky" free will invokes Cartesian dualism, demands freedom from the laws of cause and effect, and offers nothing of explanatory value in return.

Taking spooky free will off the table means we can also put to rest a persistent but misguided concern about whether or not *determinism* is true. In physics and in philosophy, determinism is the proposal that all events in the universe are completely determined by previously existing physical causes. The alternative to determinism is that chance is built into the universe from the ground up, whether through fluctuations in a quantum soup or through some other as-yet-unknown principles of physics. Whether determinism matters for free will has been the topic of endless debate. My former boss Gerald Edelman summed it up well with a provocative one-liner: Free will—whatever you think about it, we're determined to have it.

Once spooky free will is out of the picture, it is easy to see that the debate over determinism doesn't matter at all. There's no longer any need to allow any nondeterministic elbow room for it to intervene. From the perspective of free will as a perceptual experience, there is simply no need for any disruption to the causal flow of physical events. A deterministic universe can chug along just fine. And if determinism is false, it doesn't make any difference because exercising free will does not mean behaving randomly. Voluntary actions neither *feel* random, nor *are* random.

IN THE EARLY 1980S, at the University of California in San Francisco, the neuroscientist Benjamin Libet performed a series of experiments on the brain basis of voluntary action, which have remained controversial ever since. Libet took advantage of a well-known phenomenon called the "readiness potential"—a small slope-like EEG signal, originating from somewhere over the motor cortex, that reliably precedes voluntary actions. Libet wanted to know whether this brain signal could be identified not only prior to a voluntary action but before the person was even *aware of the intention* to make the action.

His experimental setup, shown in the image on page 221, was straightforward. Libet asked his participants to flex their dominant wrist at a time of their own choosing—to make a spontaneous voluntary action, just as Briony does in McEwan's novel. Each time they did this, he measured the precise time of the movement, while using EEG to record brain activity both before and after the onset of the movement. Crucially, he also asked his volunteers to estimate when they experienced the "urge" to make each movement: the precise moment of conscious intention, the crest of the breaking wave. They did this by noting the angular position of a rotating dot on an oscilloscope screen at the time they experienced the intention to move, and then reporting this position later on.

The data were clear. After averaging across many trials, the readiness potential was identifiable hundreds of milliseconds *before* the conscious intention to move. In other words, by the time a

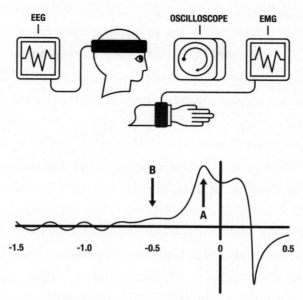

Benjamin Libet's famous volition experiment.*

person is aware of their intention, the readiness potential has already started ramping up.

A common interpretation of Libet's experiment is that it "disproves free will." Indeed, it is clearly bad news for spooky free will (not that more bad news is needed) because it seems to exclude the possibility that the experience of volition caused the voluntary action. Libet himself was sufficiently worried by this implication that, in what now seems like a desperate rescue attempt, he floated the idea that enough time remained between the moment of the

---

* A volunteer is asked to flex his wrist at a time of his choosing, while noting the position of the rotating dot on the oscilloscope at the precise time he feels the conscious intention to move. Other devices measure his muscle activity (EMG) and brain activity (EEG). The lower panel shows typical average EEG when time-locked to movement onset (0 sec). The arrows show timing of the conscious urge (A) and onset of the readiness potential (B).

urge and the resulting action for spooky free will to intervene and *prevent* the action from happening. If there isn't any genuine (i.e., spooky) free will, Libet thought, maybe there's still "free won't." This is a cute trick, but of course it doesn't work. Conscious inhibition is no more a little miracle than the original conscious intention is.

Precisely *what* Libet's observations say about free will has been debated for decades. It does seem strange that the readiness potential can be identified so long before the voluntary action. In brain time, half a second is a very long time. It wasn't until 2012 that a new idea and a clever experiment properly shook things up, when the neuroscientist Aaron Schurger realized that readiness potentials might not be signatures of the brain initiating an action, but might instead be artifacts of the way they are measured.

Readiness potentials are typically measured by looking backward in time, at the EEG, starting from all those moments at which a voluntary action actually occurred. What Schurger realized is that, by doing this, researchers systematically ignore all the other times when voluntary actions *don't* happen. What would the EEG look like at these other times? Perhaps there is activity similar to readiness potentials going on all the time, but we don't see it, because we aren't looking for it?

This reasoning can be clarified with an analogy. In the "high striker" circus game, punters swing a mallet as hard as they can, sending a small hockey puck flying upward toward a bell. If they swing hard enough, the bell rings; otherwise the puck falls back down in silence. If a circus scientist examined puck trajectories only for those occasions where the bell rang, she might mistakenly conclude that a rising puck trajectory (the readiness potential) always led to the bell ringing (the voluntary action). To understand how the high striker actually worked, she'd need also to examine puck trajectories on those occasions when the bell did *not* ring.

Schurger attacked this problem through a clever modification of the Libet design in which people continued to make spontaneous voluntary actions, but were also occasionally prompted, by a loud beep, to make the same action in a nonvoluntary, stimulus-driven way. His key finding was that when his volunteers were quick to respond to the beep, their EEG showed what looked like a readiness potential, extending back long before the beep, even though they hadn't been preparing any voluntary action at these times. By contrast, when looking at the EEG preceding slow responses to the beep, there was little sign of anything resembling a readiness potential.

Schurger interpreted his data by proposing that the readiness potential is *not* a signature of the brain initiating a voluntary action, but a fluctuating pattern of brain activity that occasionally passes a threshold, and which triggers a voluntary action when it does so. This is why, in the standard Libet experiment, you see a slowly rising slope in the EEG when you look back in time from the moments when voluntary actions happened. And this is why, when an action is triggered by a beep, the behavioral response will be quicker if this fluctuating activity happens to be close to the threshold, and slower if it happens to be far away. This in turn means that you will see something that looks like a readiness potential if you look back in time from moments of fast responses—when the activity happens to be close to threshold—but not when you look back from slow responses—when the activity is far from threshold.

Schurger's elegant experiment explains why we see readiness potentials when we look for the neural signatures of voluntary actions, and why it is misleading to think of them as being the specific causes of these actions. But then how should we interpret these fluc-tuating patterns of brain activity? My preferred interpretation re-turns to the idea I started with: that experiences of volition are forms of self-related perception. Through the lens of Schurger's ex-periment, readiness potentials look a lot like the activity associated with the brain accumulating sensory data in order to make a Bayes-ian best guess. In other words, they are the neural fingerprints of a special kind of controlled hallucination.

I just made a cup of tea.

Let's use this example to develop the view of experiences of volition—and voluntary actions too—as self-related perceptions. There are three defining features that characterize most, if not all, experiences of volition.

The first defining feature is the feeling that *I am doing what I want to do*. Being English—at least semi-English—making tea is perfectly aligned with my psychological beliefs, values, and de-sires, as well as with my physiological state at the time and the opportunities—affordances—of my environment. I was thirsty and tea was available, nobody was restraining me or force-feeding me hot chocolate, so I made some tea and drank it. (Of course, if I am being forced to do something "against my will," I may still feel my actions to be voluntary at one level, but involuntary at another.)

Although making tea was fully consistent with my beliefs, val-ues, and desires, I did not choose to have these beliefs, values, and desires. I wanted a cup of tea, but I did not choose to want a cup of tea. Voluntary actions are voluntary not because they descend from an immaterial soul, nor because they ascend from a quantum soup.

They are voluntary because they express what I, as a person, want to do, even though I cannot choose these wants. As nineteenth-century philosopher Arthur Schopenhauer put it, "Man can do what he wills, but he cannot will what he wills."

The second defining feature is the feeling that *I could have done otherwise*. When I experience an action as voluntary, the character of the experience is not only that I did X, but that I did X and not Y, even though I could have done Y.

I made tea. Could I have done otherwise? In one sense, yes. There's coffee in the kitchen too, so I could have made coffee. And when making the tea, it certainly *seemed to me* that I could have made coffee instead. But I didn't want coffee, I wanted tea, and since I can't choose my wants, I made tea. Given the precise state of the universe at the time, which includes the state of my body and brain, all of which have prior causes, whether deterministic or not, stretching all the way back to my origin as a tea-drinking semi-Englishman and beyond, *I could not have done otherwise*. You can't replay the same tape and expect a different outcome, apart from uninteresting differences due to randomness. The relevant phenomenology—the *feeling* that I could have done otherwise—is not a transparent window onto how causality operates in the physical world.

The third defining feature is that voluntary actions *seem to come from within* rather than being imposed from somewhere else. This is the difference between the experience of a reflex action, like the rapid withdrawal of my foot when I accidentally stub my toe, and its voluntary equivalent—such as when I deliberately swing my foot backward as I prepare to kick a ball. It's the feeling that Briony Tallis had as she tried to catch herself at the crest of the breaking wave of her conscious intention to flex her finger.

Altogether, we perceive an action as being voluntary—as being "freely willed"—when we infer that its causes come predominantly from within, in a way that is aligned with one's beliefs and goals, detached from alternative potential causes in the body or in the world, and that suggests the possibility of having done otherwise. This is what experiences of volition feel like from the inside, and also what voluntary actions look like from the outside.*

The next step is to ask how the brain enables and implements such actions. Here's where "degrees of freedom"—the title of this chapter—enters the story. In engineering and mathematics, a system has degrees of freedom to the extent that it has multiple ways of responding to some state of affairs. A rock has basically no degrees of freedom, whereas a train on a single track has one degree of freedom (go backward or forward). An ant might have quite a few degrees of freedom in how its biological control system responds to its environment, while you and I have vastly more degrees of freedom thanks to the spectacular complexities of our bodies and our brains.

Voluntary behavior depends on the competence to control all these degrees of freedom, in ways that are aligned with our beliefs, values, and goals, and that are adaptively detached from the immediate exigencies of the environment and body. This competence to control is implemented by the brain not by any single region where "volition" resides, but by a network of processes distributed over many regions in the brain. Execution of even the simplest voluntary action—flicking a switch to turn the kettle on, Briony flexing her finger—is underpinned by such a network. Following the neuro-scientist Patrick Haggard, we can think of this network as imple-

---

* Sometimes, voluntary actions are also experienced as requiring conscious effort, or "will-power." Writing this footnote, for example, feels effortful. But many self-initiated voluntary actions require little or no conscious effort. It is therefore important not to confuse willpower with (the experience of) free will.

menting three processes: an early "what" process specifying which action to make, a subsequent "when" process determining the timing of the action, and a late-breaking "whether" process, which allows for its last-minute cancellation or inhibition.

The "what" component of volition integrates hierarchically organized sets of beliefs, goals, and values together with perceptions of the environment, in order to specify a single action out of many possibilities. I move my hand to the kettle because I am thirsty, I like tea, it is the right time of day, the kettle is within reach, no wine is available—and so on. These nested perceptions, beliefs, and goals involve many different brain regions, with a concentration in the more frontal parts of the cortex. The "when" component specifies the timing of a selected action, and is most closely associated with the subjective urge to move—the urge that Briony Tallis wondered about and that Benjamin Libet measured. The brain basis of this process localizes to the same regions that are associated with the readiness potential. Indeed, gentle electrical brain stimulation of these regions—in particular the supplementary motor area—can generate a subjective urge to move, even in the absence of any movement. And the final "whether" component provides a last-minute check on whether the planned action should go ahead. When we call off an action at the very last moment—perhaps I'm out of milk—it's this process of "intentional inhibition" that kicks in. These inhibitory processes are also localizable to more frontal parts of the brain.

These interwoven processes play out in a continual loop spanning the brain, body, and environment, with no beginning and no end, implementing a highly flexible ongoing form of goal-oriented behavior. This network of processes funnels a large array of potential causes into a single flow of voluntary actions—and at times their inhibition. And it is the perception of the operation of this network, its looping through the body, out into the world, and back again, that underpins subjective experiences of volition.

What's more, since action itself is a form of self-fulfilling perceptual inference, as we saw in chapter 5, perceptual experiences of volition and the ability to control many degrees of freedom are two sides of the same prediction machine coin. The perceptual experience of volition is a self-fulfilling perceptual prediction, another distinctive kind of controlled—again perhaps a *controlling*—hallucination.

There's one further reason why we experience voluntary actions the way we do, a reason that puts even more clear air between volition as perceptual inference and volition as dualistic magic. Experiences of volition are useful for guiding *future* behavior, just as much as for guiding *current* behavior.

As we've seen, voluntary behavior is highly flexible. The competence to control large numbers of degrees of freedom means that if a particular voluntary action turns out badly, then the next time a similar situation arises, I might try something different. If on Monday I attempt a shortcut on my drive to work but arrive late because I get lost, then on Tuesday I may choose a longer but more reliable route. Experiences of volition flag up instances of voluntary behavior so that we can pay attention to their consequences, and adjust future behavior so as to better achieve our goals.

I mentioned earlier that our sense of free will is very much about feeling we "could have done differently." This counterfactual aspect of the experience of volition is particularly important for its future-oriented function. The feeling that I could have done differently does *not* mean that I actually could have done differently. Rather, the phenomenology of alternative possibilities is useful because in a future similar, but not identical, situation I might indeed do differently. If every circumstance is indeed identical on Tuesday as on

Monday, then I can do no differently on Tuesday than on Monday. But this will never be the case. The physical world does not duplicate itself from day to day, not even from millisecond to millisecond. At the very least, the circumstances of my brain will have changed, because I've had an experience of volition on Monday and paid attention to its consequences. This, by itself, is enough to affect how my brain can control my many degrees of freedom when setting out to work again on Tuesday.* The usefulness of feeling "I could have done otherwise" is that, next time, you might.

And who is the "you"? The "you" in question is the assemblage of self-related prior beliefs, values, goals, memories, and perceptual best guesses that collectively make up the experience of *being you*. Experiences of volition themselves can now be seen as an essential part of this bundle of selfhood—they are another species of self-related controlled, or controlling, hallucination. Altogether, the ability to exercise and to experience "free will" is the capacity to perform actions, to make choices—and to think thoughts—that are uniquely your own.

So is free will an illusion? We often hear sage pronouncements that it is. The renowned psychologist Daniel Wegner captured this spirit with his book *The Illusion of Conscious Will*, which has remained influential since its publication nearly twenty years ago. The correct answer to the question is, of course, "It depends."

Spooky free will certainly isn't real. In fact, spooky free will may not even qualify as being illusory. When examined closely, as

---

* Heraclitus: "No man ever steps in the same river twice, for it's not the same river and he's not the same man."

we've seen, the phenomenology of volition is not so much about immaterial uncaused causes, it is a self-fulfilling controlling hallucination related to specific kinds of actions—those actions that seem to come from within. Seen this way, spooky free will is an incoherent solution to a problem that doesn't exist.

And although I've concentrated in this chapter on examples in which voluntary actions are accompanied by vivid experiences of volition, this is not always the case. When I play the piano, or make a cup of tea, most of the time these voluntary actions unfold with an automaticity and fluency that undermines not only the intuition that *I* somehow cause the actions, but also the less frequently examined intuition that such actions even seem to be caused by anything. When people talk about being "in the moment" or in a "state of flow"—when deeply immersed in an activity they have extensively practiced—the phenomenology of volition may be entirely absent. Much of the time, our voluntary actions, and our thoughts—well, they "just happen." When it comes to free will, it's not only that how things seem is not how they are. How things *seem* deserves closer examination too.

From another perspective, free will is not illusory at all. So long as we have relatively undamaged brains and relatively normal upbringings, each of us has a very real capacity to execute and to inhibit voluntary action, thanks to our brain's ability to control our many degrees of freedom. This kind of freedom is both a freedom *from* and a freedom *to*. It is a freedom *from* immediate causes in the world or in the body, and from coercion by authorities, hypnotists and mesmerists, or social-media pushers. It is not, however, freedom from the laws of nature or from the causal fabric of the universe. It is a freedom *to* act according to our beliefs, values, and goals, to do as we wish to do, and to make choices according to who we are.

The reality of this kind of free will is underlined by the fact that it cannot be taken for granted. Brain injuries, or unlucky draws from the lotteries of our genes and our environment, can undermine our ability to exercise voluntary behavior. People with anarchic hand syndrome make voluntary actions that they do not experience as being theirs, while those with akinetic mutism are unable to make any voluntary actions at all. An awkwardly located brain tumor can transform an engineering student into a mass school shooter, as happened in the case of Charles Whitman, the "Texas Tower Sniper," or engender in a previously blameless teacher a rampant pedophilia—a tendency which disappeared when the tumor was removed, and returned when it grew back.

The ethical and legal quandaries raised by cases like these are also real. Charles Whitman did not choose to have the brain tumor that pressed down on his amygdala, so should he be held responsible for his actions? Intuitively, one might think not, but as we understand more about the brain basis of volition, is it not a case of "brain tumors all the way down" for each of us?* This argument works the other way around too. Einstein stated in a 1929 interview that because he didn't believe in free will, he took credit for nothing.

It is also a mistake to call the *experience of volition* an illusion. These experiences are perceptual best guesses, as real as any other kind of conscious perception, whether of the world or of the self. A conscious intention is as real as a visual experience of color. Neither

---

* Western legal systems are founded on the principle that criminal liability requires both a "guilty act" (*actus reus*) and a "guilty mind" (*mens rea*). When a person's ability to exercise free will—to control their degrees of freedom—is injured or repressed in some way, can they be said to have a "guilty mind"? Some, such as the philosopher Bruce Waller, argue that since we do not decide to have the brains that we have, the very concept of moral responsibility is incoherent. Another view, which I am attracted to, is that once we pass a certain threshold of competence to control our degrees of freedom, we *can* be held responsible for our actions.

corresponds directly to any definite property of the world—there is no "real red" or "real blue" out there, just as there is no spooky free will in here—but they both contribute in important ways to guiding our behavior, and both are constrained by prior beliefs and sensory data. Whereas color experiences construct features of the world around us, experiences of volition have the metaphysically subversive content that the "self" has causal influence in the world. We project causal power into our experiences of volition in just the same way that we project redness into our perceptions of surfaces. Knowing that this projection is going on—to channel Wittgenstein one more time—both changes everything and leaves everything just the same.

Experiences of volition are not only real, they are indispensable to our survival. They are self-fulfilling perceptual inferences that bring about voluntary actions. Without these experiences, we would not be able to navigate the complex environments in which we humans thrive, nor would we be able to learn from previous voluntary actions in order to do better the next time.

Briony Tallis thought that if she could identify the crest of volition's breaking wave, she could find herself. The self in question, of course, is a human self, and there does seem to be something distinctively human about our ability to deal with complex and changeable environments through flexible, voluntary behavior. However, the ability to exercise free will might come in degrees not only among us humans, but more widely among the animals we share our world with.

And if the ability to exercise free will extends to other species, what can be said about the extent of consciousness itself?

It's time to look beyond the human.

IV

OTHER

12

BEYOND HUMAN

FROM THE EARLY NINTH century all the way up until the mid-1700s, it was not uncommon for European ecclesiastical courts to hold animals criminally responsible for their actions. Pigs were executed or burned alive, as were bulls, horses, eels, dogs, and, on at least one occasion, dolphins. In the almost two hundred cases documented in E. P. Evans's 1906 history of animal criminal prosecution, pigs were the most common offenders, probably because they roamed rather freely in medieval villages. Their crimes varied from eating children to consuming consecrated biscuits. Sometimes they were charged with abetting—through their grunting and snorting—the crimes of another; often they were hanged, and occasionally they were acquitted.

Plagues of rodents, locusts, weevils, and other such smaller animals were less easy to deal with via legal proceedings. In one celebrated sixteenth-century case, the French lawyer Bartholomew Chassenée successfully exonerated some rats with the clever argument that they could not reasonably be expected to turn up to trial, given the dangers posed to them by the many cats lying in wait along the route. In other cases, including various weevil infestations, the

A sow and her piglets are tried for the murder of a child (possibly from 1457).

offending animals were issued with written orders to leave a property or a barley crop, often on a specific day and even by a specific hour.

As bizarre as all this seems to our twenty-first-century mindset, the medieval perspective on animal minds foreshadowed the recent resurgence of interest in animal consciousness, and in whether "personhood" can extend beyond the human.* The idea that animals could comprehend, and reasonably submit to, the arcane procedures of ecclesiastical law was and is borderline insane. But along with this idea comes a recognition that animals might have conscious experiences, and might be equipped with minds able—in some sense—to make decisions. This recognition of conscious minds beyond the human stands in stark contrast to the Cartesian version of the beast machine story, in which animals lack the conscious status

---

* I will use the word "animal" as a shorthand for "nonhuman" animals. Humans are animals too.

that goes along with rational minds. Animals, for many medievals, were certainly beasts. But they were not the animal-robots of Cartesian dualism. They had their inner universes too.

These days, it would be strange and almost perverse to argue that only humans are conscious. But what can we really say about how far the circle of consciousness extends, and about how different the inner universes of other animals might be?

THE FIRST THING TO SAY is that we cannot judge whether an animal is conscious by its ability—or inability—to tell us that it is conscious. Absence of language is not evidence for absence of consciousness. Neither is absence of so-called "high-level" cognitive abilities like metacognition—which is the ability, broadly speaking, to reflect on one's thoughts and perceptions.

Animal consciousness, where it exists, will be different—and in some cases very different—from our own. Although animal experiments can shed light on the mechanisms of consciousness in humans, it is unwise to infer the existence of animal consciousness solely on the basis of superficial similarity to *Homo sapiens*. Doing so carries the twin risks of *anthropomorphism*—the attribution of humanlike qualities to the nonhuman—and *anthropocentrism*—the tendency to interpret the world in terms of human values and experiences. Anthropomorphism encourages us to see humanlike consciousness where it might not be—such as when we believe our pet dog really understands what we are thinking. Anthropocentrism, on the other hand, blinds us to the diversity of animal minds, preventing us from recognizing non-humanlike consciousness where it might actually be—a myopia exemplified by the Cartesian view of animals as beast machines.

Above all, we should be suspicious of associating *consciousness* too closely with *intelligence*. Consciousness and intelligence are not the same thing. Using the latter as a litmus test for the former commits a number of errors. It falls foul of anthropocentrism: human beings are intelligent and conscious, therefore for animal X to be conscious, it must also be intelligent. It falls foul of anthropomorphism too: we see humanlike intelligence in animal X, but not in animal Y, therefore animal X but not animal Y is conscious. And it encourages a methodological laziness, since it justifies accepting "intelligent" capabilities like language and metacognition—which are easier to assess than consciousness itself—as sufficient for inferring consciousness.

Intelligence is not irrelevant to consciousness. Other things being equal, intelligence opens up new possibilities for conscious experience. You can be sad or disappointed without much cognitive competence at all, but to feel regret—or anticipatory regret—requires enough mental capability to consider alternative outcomes and courses of action. Even rats, one study suggests, might experience a rodent version of regret rather than just disappointment when things don't turn out as hoped for.*

Inferences about nonhuman consciousness must tread a fine line. We need to be wary of imposing our anthropocentric view, but at the same time we have little option but to use humans as a known quantity; a firm foundation from which to reach outward. After all, we know that we are conscious, and we have an increasing grasp on

---

* In this study, conducted by Adam Steiner and David Redish, rats had to decide between different options which were associated with different levels of reward. When they chose an option which delivered less reward than expected, they were more likely to look back at the option not chosen. The researchers interpreted this as "expressing something akin to human regret," though it is far from clear what—if anything—the rats were actually feeling.

the brain and bodily mechanisms involved in human consciousness which we can use as a basis for extrapolation.

The beast machine theory developed in this book makes the case that consciousness is more closely connected with being *alive* than with being *intelligent*. Naturally, this applies as much to other animals as it does to us humans. In this view, consciousness may be more widespread than it would seem, were we to take intelligence as the primary criterion. But it does not mean that wherever there is life there is also consciousness.

Looking for consciousness beyond the human is like stepping out onto an icy lake from a frozen shore. One careful step at a time, always checking how solid the ice feels beneath your feet.

LET'S START WITH MAMMALS—a grouping which includes rats, bats, monkeys, manatees, lions, hippos, and of course humans too. I believe that *all mammals are conscious*. Of course, I don't know this for sure, but I am pretty confident. This claim is not based on superficial similarity to humans, but on shared mechanisms. If you set aside raw brain size—which has more to do with body size than with anything else—mammalian brains are strikingly similar across species.

Back in 2005, the cognitive scientist Bernard Baars, myself, and David Edelman—the son of Gerald Edelman, and an expert on animal cognition—made a list of the properties of human consciousness which we thought could be readily tested for in other mammals. We came up with seventeen distinct properties. This was in some ways an arbitrary number, but it demonstrated the reasonableness of asking experimentally testable questions about animal consciousness.

The first properties we thought about had to do with anatomical features of the brain. In terms of brain wiring, the primary neuro-anatomical features that are strongly associated with human consciousness are found in all mammalian species. There is a six-layered cortex, a thalamus that is strongly interconnected with this cortex, a deep-lying brainstem, and a host of other shared features—including neurotransmitter systems—which are consistently implicated, in humans, in the moment-to-moment flow of conscious experience.

There are common features of brain activity too. Among the most striking are the changes in brain dynamics as animals fall asleep and wake up—the dynamics underlying conscious *level*. In normal waking states, all mammals show irregular, low-amplitude, and fast electrical brain activity. And when sleep comes, all mammalian brains switch to more regular, large-amplitude brain dynamics. These patterns and changes closely resemble what is seen in humans during waking and sleeping. General anesthesia, too, has similar effects across mammalian species—a widespread breakdown of communication between brain regions that accompanies total behavioral unresponsiveness.

There are of course many differences as well, especially in patterns of sleep. Seals and dolphins sleep with half their brain at a time, koalas sleep for about twenty-two hours each day, while giraffes get by on less than four, and newborn killer whales do not sleep at all in the first month of life. Nearly all mammals have periods of rapid eye movement (REM) sleep—though seals do so only while sleeping on land, and dolphins apparently not at all.

Besides conscious level, there will also be substantial differences in conscious *contents* across mammalian species. Much of this variation can be attributed to differences in dominant kinds of perception. Mice rely on their whiskers, bats on their echolocating sonar, and naked mole rats on their acute sense of smell—especially when meeting other naked mole rats. These differences in perceptual dominance will mean that each animal will inhabit a distinctive inner universe.*

More intriguing still are differences that relate to experiences of *selfhood*. In humans, a notable marker of the development of high-level self-consciousness, of the sort related to personal identity, is the ability to recognize oneself in the mirror. In humans, this "mirror self-recognition" ability tends to develop some time between eighteen and twenty-four months of age. This doesn't mean that younger infants lack consciousness—only that their awareness of themselves as an individual, as separate from others, may not have fully formed before this age.

In animals, the capacity for self-recognition has been extensively investigated using a test developed by the psychologist Gordon Gallup Jr. in the 1970s. In the classic version of his mirror self-recognition test, an animal is anesthetized and then marked, usually with paint or a sticker, in a place on its body that it cannot normally see. When the anesthesia wears off, the animal is allowed to interact with a mirror so that it can see the mark. If, having looked into the mirror, it spontaneously looks for the mark on its own body, instead of investigating the mirror image, it has passed the test. This criterion is based on the reasoning that the creature has recognized that the mirror image depicts its own body, rather than the body of another animal.

Who passes the mirror test? Among mammals, some great apes, a few dolphins and killer whales, and a single Eurasian elephant. A

---

* The world as experienced by an animal is often called the *Umwelt* for that animal—a term introduced by the ethologist Jakob von Uexküll.

parade of other mammalian creatures, including pandas, dogs, and various monkeys, have failed—at least so far. Given how intuitive mirror self-recognition is for us humans, and how otherwise cognitively competent many of these non-self-recognizing mammals seem to be, this pass list is remarkably short. There is no convincing evidence that any non-mammal passes the mirror test, although manta rays and magpies may come close, and there is currently some controversy about the cleaner wrasse.

Animals could fail the mirror test for many reasons besides lacking self-recognition abilities. These include not liking mirrors, not understanding how they work, or even just preferring to avoid eye contact. Recognizing this, researchers are continually developing new versions of the test that are tuned ever more astutely to different interior universes—different perceptual worlds. For example, dog self-recognition can now be tested with "olfactory mirrors"—though they still don't do very well (delightfully, cognition in dogs is known as "dognition"). It's possible that as experimental ingenuity continues to develop, species currently on one side of the line will cross over into the light of mirror-certified self-awareness. But even if they do, the diversity of different mirror tests—and the inability of many animals to pass even heavily species-adapted tests—suggests the likelihood of dramatic differences in how mammals experience "being themselves."

THESE DIFFERENCES STRIKE me especially forcefully in the case of monkeys. Although chimpanzees and the great apes are our nearest evolutionary neighbors, monkeys are not too far away and they've long been used in neuroscientific experiments as "primate models" of humans, especially when it comes to vision. In some studies, monkeys have even been trained to deliver "reports"—for example, by pressing a lever—about whether they "saw" something or not. These experiments can be directly compared with human studies in which people say what they experience, or don't experience, providing a primate equivalent of a key method in consciousness research.

Given their many similarities to humans, for me there is no doubt that monkeys possess some kind of conscious selfhood. If you hang around with monkeys for any length of time, the impression of being among other conscious entities—other conscious *selves*—is completely convincing.

I experienced this in July 2017 when spending a day on Cayo Santiago, a small island just off the eastern coast of Puerto Rico, in the Caribbean. Cayo Santiago is also known as "monkey island" because its only permanent residents are rhesus macaque monkeys—more than a thousand of them. The population was transplanted there in 1938 from Kolkata by an eccentric American zoologist, Clarence Ray Carpenter, who had become tired of trekking all the way to India. As I wandered around Cayo Santiago on this hot summer day, in the company of the Yale psychologist Laurie Santos (and a film crew), dozens of monkeys went about their business, wary but tolerant of us lumbering humans. When two monkeys took turns to climb a tree and leap from its branches into a pond below, it seemed to be for no other reason than the sheer spontaneous pleasure of it. They were having *fun*.*

Equally compelling are videos of capuchin monkeys reacting

---

* Shortly after our visit, Cayo Santiago—along with much of Puerto Rico—was devastated by Hurricane Maria. Thankfully, most of the monkeys survived. Much of the research infrastructure, however, was destroyed. The footage we recorded features in the 2018 documentary *The Most Unknown* (www.themostunknown.com).

to deliberate unfairness. In one video, popularized by the prima-
tologist Frans de Waal, two monkeys are housed in adjacent cages
and are rewarded, one after the other, for passing a small stone to
an experimenter. Monkey One passes the stone through the cage
mesh and is rewarded with a small slice of cucumber, which she
happily eats. Monkey Two does the same thing and is given, not
cucumber, but a grape—a much tastier morsel. Monkey Two eats
the grape while Monkey One looks on. When Monkey One repeats the task and is again given cucumber, she looks at it, throws
it back at the experimenter, and rattles the cage in apparent indig-
nation.

Having fun and throwing tantrums are powerful intuition
pumps. These behaviors are so distinctive that it's almost impossible
to interpret them as anything other than the outward manifestation
of apparently humanlike inner states. When we witness a monkey
behaving like this, we intuit the presence not just of another conscious being, but of a conscious being *like us*. But here's the thing.
Monkeys, as mentioned, have consistently failed the mirror test.
While monkeys are undoubtedly conscious, and while I also believe
they experience some kind of selfhood, they are not furry little
people.

The shaping of our intuitions by anthropomorphism and anthropocentrism becomes even more apparent when we look beyond
mammals. Especially when we look as far as some of our most distant evolutionary relatives.

IN THE SUMMER OF 2009 David Edelman and I spent a week with
about a dozen *Octopus vulgaris*—the common octopus, a species of

cephalopod.* We were visiting the biologist Graziano Fiorito, a leading expert on cephalopod cognition and neurobiology. Although
more than a decade has since passed, this week still stands as one of
the most memorable in all my years as a scientist.

Fiorito's octopus lab—part of the renowned Italian research institute the Stazione Zoologica—is located in a dank basement directly beneath a public aquarium in the heart of Naples, a cool
refuge from the raucous summer heat above. My week there was
taken up mainly by spending time with these fascinating creatures,
observing how they change shape, color, and texture, and paying
attention to what they were paying attention to. One day, as I was
trying to match the ever-changing appearance of one particular octopus to the drawings in Fiorito's *A Catalog of Body Patterning in
Cephalopoda*, I heard a dull splat and a slither. I'd left a tank lid ajar
and the creature was making a break for it. To this day I am convinced that it had lulled me into a false sense of security, biding its
time until I turned away for a moment too long.

While I was causing havoc, David was putting together an experiment on visual perception and learning. He would lower differently shaped objects into the tank of an octopus, some of which
were accompanied by a tasty crab. The idea was to see whether the
octopus could learn to associate particular objects with reward. I
don't remember exactly how the study turned out, but I do remember one episode very clearly.

Fiorito's lab was arranged with two rows of tanks lining a central

---

* Cephalopods include octopuses, squid, and cuttlefish, as well as comparatively simple creatures like nautilus, among roughly eight hundred extant species. The term "cephalopod"
translates literally as "head-foot," which is unfortunate for octopuses, which have armlike
appendages rather than feet attached to the head.

walkway, one octopus in each tank. (Octopuses are generally not so-cial creatures and can even be cannibalistic.) On this particular day, David had chosen a tank in the left-hand row, about halfway along. When I walked in to see what was going on, I was astonished to see all the octopuses on the other side of the walkway pressed up against the glass of their tanks, every one of them staring intently at David while he repeatedly lowered his objects into his chosen tank. The observing octopuses seemed to be trying to figure out what was going on for no other reason than the sheer interest of it.

Being among octopuses, even for a short time, left me with an impression of an intelligence, and a conscious presence, very differ-ent from any other—and certainly very different from our own human incarnation. This of course was a subjective impression, nec-essarily tainted by the biases of anthropomorphism and anthropo-centrism, and open to the charge of taking intelligence as a sign of sentience. But the octopus is objectively remarkable too, and spend-ing some time with them can push our intuitions about how differ-ent a nonhuman consciousness might be.

The most recent common ancestor of humans and octopuses lived about 600 million years ago. Little is known about this ancient crea-ture. Perhaps it was some kind of flattened worm. Whatever it looked like, it must have been a very simple animal. Octopus minds are not aquatic spin-offs from our own, or indeed from any other species with a backbone, past or present. The mind of an octopus is an inde-pendently created evolutionary experiment, as close to the mind of an alien as we are likely to encounter on this planet. As scuba-diving philosopher Peter Godfrey-Smith put it, "If we want to understand *other* minds, the minds of cephalopods are the most other of all."

The body of an octopus is remarkable enough. The common octopus, *O. vulgaris*, has eight armlike appendages, three hearts pumping blue blood, an ink-based defense mechanism, and highly developed jet propulsion. An octopus can change size, shape, texture, and color at will, and all at the same time if necessary. It is a liquid animal: apart from a centrally located bony beak, the octopus is entirely soft-bodied, allowing it to squeeze through unfeasibly tiny gaps—as I discovered for myself at the Stazione Zoologica.

These extraordinary bodies are complemented by highly sophisticated nervous systems. *Octopus vulgaris* has about half a billion neurons, roughly six times more than a mouse. Unlike in mammals, most of these neurons—about three-fifths—are in its arms rather than in its central brain, a brain which nonetheless boasts about forty anatomically distinct lobes. Also unusual is that octopus brains lack myelin—the insulating material that in mammalian brains helps long-range neural connections develop and function. The octopus nervous system is therefore more distributed and less integrated than mammalian nervous systems of similar size and complexity. Octopus consciousness—assuming there is such a thing—may therefore also be more distributed and less integrated, perhaps even without having a single "center" at all.

Octopuses do things differently even at the level of genes. In most organisms, genetic information in DNA is transcribed directly into shorter sequences of RNA (ribonucleic acid) which are then used to make proteins—the molecular workhorses of life. This is a well-established, textbook-level principle of molecular biology. But in 2017, this principle was upended by the discovery that RNA sequences in octopuses—and in a few other cephalopods—can undergo significant editing before being translated into protein. It's as if the octopus is able to rewrite parts of its own genome on the fly. (RNA editing had been previously identified in other species, but in those instances, it plays only a relatively minor role.) What's more, for the octopus, much of this RNA editing seems to be related to the nervous system. Some researchers have suggested that this prolific genome rewriting ability may partly underlie the impressive cognitive abilities of octopuses.

And their cognitive abilities certainly are impressive. They can retrieve hidden objects—usually tasty crabs—from within nested Plexiglas cubes, find their way through complex mazes, try out a range of different actions to solve a particular problem, and—as Fiorito himself has shown at the Stazione—learn simply through observing other octopuses. Anecdotal reports of octopus behavior in the wild are even more astonishing. In one of the more extraordinary examples, footage from the BBC television series *Blue Planet II* shows an octopus caught out in the open, covering itself with shells and other seafloor detritus in order to hide from a predatory shark.

These feats of cephalopod intelligence are certainly compelling evidence of a mind at work. But what kind of mind? I've already said that we should not put too much weight on intelligence as a benchmark for consciousness. So what can be said about what it is like to be an octopus? To get a handle on this, we need to connect octopus behavior to octopus perception.

CAMOUFLAGE IS POSSIBLY THE MOST otherworldly entry in the catalog of cephalopod abilities. Without hard shells for protection, their survival often depends on their ability to meld into the background. They can match the color, shape, and texture of their surroundings so completely that you or I, along with many potential predators, would find the animal essentially invisible, even from just a meter or two away.

Octopuses color-match with their surroundings by making use

of an exquisitely precise system of *chromatophores*. These are small elastic sacs, distributed all over the skin, which produce red, yellow, or brown coloration when they are opened by neural commands originating mainly from chromatophore lobes in the brain. Exactly how this works is still not fully understood. Part of the challenge is that octopuses must render themselves invisible not to other octopuses, but to predators who see the world in their own distinctive ways. Their camouflage system must therefore somehow encode knowledge about the visual abilities of these predators.

What's even more surprising is that octopuses do all this despite being color-blind. The light-sensitive cells in human eyes respond to three different wavelengths of light, creating from their mixtures a universe of color. The cells in octopus eyes, however, contain only one photopigment. Octopuses can sense the direction of polarization of light—just like you or I can when wearing polarizing sunglasses—but they cannot conjure colors out of combinations of wavelength. The same color blindness is true for the light-sensitive cells embedded throughout their skin: it turns out that octopuses can "see" with their skin, as well as with their eyes. Added to this, octopus chromatophore control is thought to be "open-loop," meaning that the neurons in the chromatophore lobe do not generate any obvious internal copy of the signals sent out to the chromatophores in the skin. The central brain might not even know what its skin is doing.

It's hard to wrap one's head around what this means for how an octopus might experience its world, and its body within that world. Its own skin will change color in ways that it cannot itself see and which are not even relayed to its brain. And some of these adaptations may happen through purely local control, in which an arm senses its own immediate environment and changes its appearance without the central brain ever getting involved. The human-centered assumption that we can see and feel what's happening to our own bodies just doesn't apply. It's not surprising that octopuses show no signs of passing the mirror test.

Along with vision, octopuses share some of the other classic sensory modalities with mammals and other vertebrates. They can taste, smell, and touch—and they can hear too, though not very well. There's still bizarreness to contend with, because octopuses can taste with their suckers as well as with their central mouth parts. This again points to a remarkable decentralization of mind in these creatures.

The idea of a decentralized consciousness is particularly challenging when it comes to experiences of body ownership. As we saw in chapter 8, in humans this aspect of conscious selfhood can be altered surprisingly easily, by tricking the brain into changing its Bayesian best guess about what is, and what is not, part of the body. Keeping track of the body is hard enough for us humans with our four limbs constrained by just a few joints. For an octopus, with its eight highly flexible arms furling and unfurling in several directions at once, the challenge is formidable. And just as sensation is partly devolved to these arms, so is control. Octopus arms behave like semiautonomous animals: a severed octopus arm can still execute complex action sequences, like grasping pieces of food, for some time after being separated from the body.

These degrees of freedom and decentralized control together pose an intimidating challenge for any central brain trying to maintain single, unified perception of what is, and what is not, part of its body. Which is why octopuses may not even bother. As odd as it sounds, what it is like to be an octopus may not include an experience of body ownership in anything like the sense in which it applies to humans and other mammals.

This doesn't mean that octopuses do not distinguish "self" from "other." They clearly do—and they need to. For a start, they need to avoid getting tangled up with themselves. The suckers on an octopus arm will reflexively grip on to almost any passing object, yet they will not grip on to other arms from the same octopus, nor on to its central body. This demonstrates that octopuses are able, in some way, to discriminate what is their body from what is not.

It turns out that this ability depends on a simple but effective system of taste-based self-recognition. Octopuses secrete a distinctive chemical throughout their skin. This chemical serves as a signal that can be detected by the suckers, so that they do not reflexively attach. In this way, an octopus can tell *what* in the world is part of itself, and what is not, even though it doesn't necessarily know *where* its body is in space. This discovery was established in a series of admittedly macabre experiments in which researchers offered detached octopus arms other detached arms, either with the skin on or with the skin removed. The detached arms would readily grip on to the arms from which the skin had been removed, but would never grip on to the intact arms.*

What this means for experiences of embodiment in an octopus is hard for us mammals to imagine. The octopus as a whole might have only a hazy perception of the what and where of its body, though it would probably not experience this perception as being hazy. And there might even be something it is like to be an octopus arm.

OCTOPUSES PUSH HARD at our intuitions about how different animal consciousness might be from our own. But in leaping straight from

---

* These experiments are not quite as awful as they sound. Octopuses do not seem to notice much when an arm is removed, and the severed arms grow back quite quickly. Of course, this does not mean that you should do experiments like these without extremely good reason.

monkeys to cephalopods, we've skipped over an enormous menag-
erie. Away from the safe shores of mammalian consciousness there
lies a vast expanse of potential animal awareness, ranging from parrots to single-celled paramecia. Contemplating this terrain, let's return to the more fundamental question of which animals are likely to
have any kind of conscious experience at all—those animals for
which the "lights are on," even if the light is just a glimmer.

Birds make a pretty strong case for sentience. Avian brains,
while significantly different from mammalian brains, nevertheless
have an organization which can be mapped quite closely onto the
mammalian cortex and thalamus. Many bird species are also strikingly intelligent. Parrots can count, cockatoos can dance, and scrub
jays can stash food according to their future needs. While these
examples of smartness suggest that some birds may enjoy complicated states of consciousness, remember that intelligence is not a
litmus test for awareness. Non-food-hiding, non-speaking, non-
dancing birds likely have conscious experiences too.

As we move further out, the evidence becomes sparser and
sketchier, and inferences about consciousness more tentative. Instead of basing these inferences on similarity to mammalian brains
and behaviors, a better strategy might be to adopt the beast machine perspective—mine, not the Cartesian version—which traces
the origin and function of conscious perception to physiological
regulation, and to preservation of the integrity of the organism.
This suggests that one place to look for evidence of awareness is in
how animals respond to supposedly painful events.

This strategy is not only scientifically sensible, it is also motivated
by ethical imperatives. Decisions about animal welfare should be
based not on similarity to humans, nor on whether some arbitrary
threshold of cognitive competence is exceeded, but on the capacity

for pain and suffering. And while there are infinitely many ways in which living creatures may suffer, the most widely shared likely involve basic challenges to their physiological integrity.

To the extent that it has been looked for, there is widespread evidence for adaptive responses to painful events among animal species. Most vertebrates (animals with backbones) will tend to an injured body part. Even the tiny zebrafish will pay a "cost" to access pain relief upon injury, shifting from a natural environment to a barren, brightly lit tank when that tank is suffused with analgesia. Whether this implies that fish are conscious—and there are *many* types of fish—is unclear, but it is certainly suggestive.

What about insects? Ants do not limp when a leg is damaged. However, their hard-bodied exoskeletons might be less susceptible to pain, and insect brains do possess forms of the opiate neurotransmitter system that is commonly associated with pain relief in other animals. A recent study found that the fruit fly *Drosophila melanogaster* displays a post-injury hypersensitivity to previously non-painful stimuli in a way that resembles "chronic pain" in humans. And, remarkably, anesthetic drugs seem to be effective across *all* animals, from single-celled critters all the way to advanced primates.

All of this is suggestive, none of it is conclusive.

At some point, it becomes difficult to say anything substantive. My intuition—and it is no more than an intuition—is that there will be some animals which do not participate in the circle of consciousness at all. One reason I feel this way is because even in mammals, with our complex brains and finely honed perceptual systems geared toward preserving physiological integrity, unconsciousness remains rather easy to achieve. Conscious experience is central to our lives, but this doesn't mean its biological basis is straightforward. By the time we reach the nematode worm with its paltry 302 neurons, I find it difficult to ascribe any meaningful conscious status—and the single-celled paramecium just doesn't make the grade.

SETTING ASIDE ITS INEVITABLE UNCERTAINTIES, the study of animal consciousness delivers two profound benefits. The first is a recognition that the way we humans experience the world and self is not the only way. We inhabit a tiny region in a vast space of possible conscious minds, and the scientific investigation of this space so far amounts to little more than casting a few flares out into the darkness. The second is a newfound humility. Looking out across the wild diversity of life on Earth, we may value more—and take for granted less—the richness of subjective experience in all its variety and distinctiveness, in ourselves and in other animals too. And we may also find renewed motivation to minimize suffering wherever, and however, it might appear.

I started this chapter by arguing that consciousness and intelligence are not the same thing, and that consciousness has more to do with being alive than with being smart. I want to finish with an even stronger claim. Not only can consciousness exist without all that much intelligence—you don't have to be smart to suffer—but intelligence can exist without consciousness too.

The possibility of being smart without suffering brings us to the final leg of our journey through consciousness science. It's time to talk about artificial intelligence, and about whether there could ever be conscious machines.

13

MACHINE MINDS

In Prague, in the late sixteenth century, Rabbi Judah Loew ben Bezalel took clay from the banks of the Vltava River and from this clay shaped a humanlike figure—a golem. This golem—which was called Josef, or Yoselle—was created to defend the rabbi's people from anti-Semitic pogroms, and apparently did so very effectively. Once activated by magical incantation, golems like Josef could move, were aware, and would obey. But with Josef, something went terribly wrong and its behavior changed from lumpen obedience into violent monstering. Eventually the rabbi managed to revoke his spell, upon which his golem fell into pieces in the synagogue grounds. Some say its remains lie hidden in Prague to this day, perhaps in a graveyard, perhaps in an attic, perhaps waiting, patiently, to be reactivated.

Rabbi Loew's golem reminds us of the hubris we invite when attempting to fashion intelligent, sentient creatures—creatures in the image of ourselves, or from the mind of God. It rarely goes well. From the monster in Mary Shelley's *Frankenstein* to Ava in Alex Garland's *Ex Machina*, by way of Karel Čapek's eponymous robots, James Cameron's *Terminator*, Ridley Scott's replicants in *Blade*

Runner, and Stanley Kubrick's HAL, these creations almost always turn on their creators, leaving in their wake trails of destruction, melancholy, and philosophical confusion.

Over the last decade or so, the rapid rise of AI has lent a new urgency to questions about machine consciousness. AI is now all around us, built into our phones, our fridges, and our cars, powered in many cases by neural network algorithms inspired by the architecture of the brain. We rightly worry about the impact of this new technology. Will it take away our jobs? Will it dismantle the fabric of our societies? Ultimately, will it destroy us all—whether through its own nascent self-interest or because of a lack of programming foresight which leads to the Earth's entire resources being transformed into a vast mound of paper clips? Running beneath many of these worries, especially the more existential and apocalyptic, is the assumption that AI will—at some point in its accelerating development—become conscious. This is the myth of the golem made silicon.

What would it take for a machine to be conscious? What would the implications be? And how, indeed, could we even distinguish a conscious machine from its zombie equivalent?

WHY MIGHT WE EVEN *THINK* that a machine—an artificial intelligence—could become aware? As I just mentioned, it is quite common—though by no means universal—to think that consciousness will emerge naturally once machines pass some as-yet-unknown threshold of intelligence. But what drives this intuition? I think that two key assumptions are responsible, and neither is justifiable. The first assumption is about the *necessary* conditions for anything to be conscious. The second is about what is *sufficient* for a specific thing to be conscious.

The first assumption—the necessary condition—is *functionalism*. Functionalism says that consciousness doesn't depend on what a system is made out of, whether wetware or hardware, whether neurons or silicon logic gates—or clay from the Vltava River. Functionalism says that what matters for consciousness is what a system *does*. If a system transforms inputs into outputs in the right way, there will be consciousness. As I explained in chapter 1, there are two separate claims here. The first is about independence from any particular substrate or material, while the second is about the sufficiency of input-output relations. Most of the time they go together, but sometimes they can come apart.

Functionalism is a popular view among philosophers of mind, and is often accepted as a default position by many nonphilosophers too. But this does not mean it is correct. For me, there are no knockdown arguments either for or against the position that consciousness is substrate-independent, or that it is solely a matter of input-output relations, of "information processing." My attitude toward functionalism is one of suspicious agnosticism.

For artificially intelligent computers to become conscious, functionalism would have to be true. This is the necessary condition. But functionalism being true is, by itself, not enough: information processing by itself is not sufficient for consciousness. The second assumption is that the kind of information processing that is sufficient for consciousness is also that which underpins intelligence. This is the assumption that consciousness and intelligence are intimately, even constitutively, linked: that consciousness will just come along for the ride.

But this assumption is also poorly supported. As we saw in the previous chapter, the tendency to conflate consciousness with intelligence traces to a pernicious anthropocentrism by which we overinterpret the world through the distorting lenses of our own values and experiences. *We* are conscious, *we* are intelligent, and we are so species-proud of our self-declared intelligence that we assume that intelligence is inextricably linked with our conscious status and vice versa.

Although intelligence offers a rich menu of ramified conscious states for conscious organisms, it is a mistake to assume that intelligence—at least in advanced forms—is either necessary or sufficient for consciousness. If we persist in assuming that consciousness is intrinsically tied to intelligence, we may be too eager to attribute consciousness to artificial systems that appear to be intelligent, and too quick to deny it to other systems—such as other animals—that fail to match up to our questionable human standards of cognitive competence.

Over the last few years, these assumptions about necessity and sufficiency have been dressed up and pushed out the door by a host of other concerns and misapprehensions, giving the prospect of artificial consciousness an urgency and an apocalyptic gloss that it doesn't really deserve.

Here are some of them. There is the worry that AI—whether conscious or not—is on a runaway path to overtake human intelligence, bootstrapping itself beyond our comprehension and our control. This is the so-called "Singularity" hypothesis, popularized by the futurist Ray Kurzweil and motivated by the extraordinary growth in raw computational resources over the last few decades. Where are we on this exponential curve? The problem with exponential curves—as many of us learned during the recent coronavirus pandemic—is that wherever you stand on them, what's ahead looks impossibly steep and what's behind looks irrelevantly flat. The local view gives no clue to where you are. Then there are our Promethean fears that our creations will turn on us in some way or another—fears which have been recognized, repackaged, and sold back to us by any number of science fiction movies and books. Finally, there is the unfortunate fact that the term "consciousness" is often bandied about with an unhelpful sloppiness when it comes to the capabilities of machines. For some people—including some AI researchers—anything that responds to stimulation, that learns something, or that behaves so as to maximize a reward or achieve a goal is conscious. To me this is a nonsensical overextension of what "being conscious" reasonably means.

Mix all these ingredients together and it is hardly surprising that many people think that conscious AI is just around the corner, and that we should be very worried about what happens when it arrives. The possibility cannot be ruled out completely. If the Singularity-mongers do turn out to be right, then we should indeed be worried. But from where we stand now, the prospect is extremely unlikely. Much more likely is a situation of the sort illustrated on page 260. Here, consciousness is not determined by intelligence, and intelligence can exist without consciousness. Both come in many forms and both are expressed along many different dimensions—meaning that there is not one single scale for either consciousness or intelligence.

In this depiction, you'll notice that current AI is located quite low on the intelligence scale. This is because it's unclear whether current AI systems are intelligent in any meaningful sense. Much of today's AI is best described as sophisticated machine-based pattern recognition, perhaps spiced up with a bit of planning. Whether intelligent or not, these systems do what they do without being conscious of anything.

Projecting into the future, the stated moonshot goal of many AI

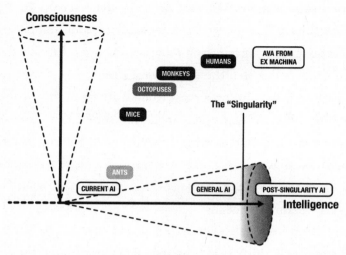

Consciousness and intelligence are separable and multidimensional. The positions of animals and machines (real and imaginary) are illustrative.

researchers is to develop systems with the general intelligence capabilities of a human being—so-called "artificial general intelligence," or "general AI." And beyond this point lies the terra incognita of post-Singularity intelligence. But at no point in this journey is it warranted to assume that consciousness just comes along for the ride. What's more, there may be many forms of intelligence that deviate from the humanlike, complementing rather than substituting or amplifying our species-specific cognitive tool kit—again without consciousness being involved.

It may turn out that some specific forms of intelligence are impossible without consciousness, but even if this is so, it doesn't mean that *all* forms of intelligence—once exceeding some as yet unknown threshold—require consciousness. Conversely, it could be that all conscious entities are at least a little bit intelligent, if intelligence is defined sufficiently broadly. Again, this doesn't validate intelligence as the royal road to consciousness.

Just making computers smarter is not going to make them sentient. But this does not mean that machine consciousness is impossible. What if we were to try to design in consciousness from the outset? If not intelligence, what *would* it take to build a conscious machine?

ANSWERING THIS QUESTION depends on what you think is sufficient for a system to be conscious, and this depends on which theory of consciousness you subscribe to. It is therefore not surprising that there are many views about what it would take for a machine to be conscious.

At the more liberal end of the spectrum are those who believe, in line with functionalism, that consciousness is simply a matter of the right kind of information processing. This information processing need not be identical with "intelligence," but it is information processing nonetheless, and therefore is the sort of thing that can be implemented in computers. For example, according to a proposal in the journal *Science* in 2017, a machine could be said to be conscious if it processes information in ways that involve "global availability" of the information, and that allow "self-monitoring" of its performance. The authors equivocate about whether such a machine would *actually be* conscious or merely *behave as if* it were conscious, but the underlying claim is that nothing more is needed for consciousness than information processing of the right kind.

A stronger claim about conscious machines has been made by advocates of integrated information theory (IIT). As we saw in chapter 3, IIT claims that consciousness simply *is* integrated information, and that the amount of integrated information produced by a system is fully determined by properties of its internal mechanisms—by its "cause-effect structure." According to IIT, any machine that generates integrated information, whatever it is made out of, and no matter what it might look like from the outside, will have some degree of consciousness. However, IIT also leaves open the possibility that machines might appear to an external observer to be conscious, or intelligent, or perhaps both, but have mechanisms that generate no integrated information at all, and therefore be lacking in consciousness altogether.

Neither of these theories identifies consciousness with intelligence, but both allow that machines satisfying some specific conditions—information processing of the right kind, or nonzero integrated information—would be conscious. But to accept these implications, it is of course necessary to accept the theories too.

THE BEAST MACHINE THEORY grounds experiences of world and self in a biological drive toward staying alive. What does this theory say about the possibility of conscious machines?

Imagine a near-future robot with a silicon brain and a human-like body, equipped with all kinds of sensors and effectors. This robot is controlled by an artificial neural network designed according to the principles of predictive processing and active inference. The signals flowing through its circuits implement a generative model of its environment, and of its own body. It is constantly using this model to make Bayesian best guesses about the causes of its sensory inputs. These synthetic controlled (and controlling) hallucinations are geared, by design, toward keeping the robot in an optimal functional state—to keep it, by its own lights, "alive." It even has artificial interoceptive inputs, signaling its battery levels and the integrity of its actuators and synthetic muscles. Control-oriented best guesses about these interoceptive inputs generate synthetic emotional states that motivate and guide its behavior.

This robot behaves autonomously, doing the right thing at the right time to fulfill its goals. In doing so, it gives the outward impression of being an intelligent, sentient agent. Internally, its mechanisms map directly onto the predictive machinery which I've suggested underlies basic human experiences of embodiment and selfhood. It is a silicon beast machine.

Would such a robot be conscious?

The unsatisfying but honest answer is that I don't know for sure, but probably not. The beast machine theory proposes that consciousness in humans and other animals arose in evolution, emerges in each of us during development, and operates from moment to moment in ways intimately connected with our status as living systems. All of our experiences and perceptions stem from our nature as self-sustaining living machines that care about their own persistence. My intuition—and again it's only an intuition—is that the materiality of life will turn out to be important for all manifestations of consciousness. One reason for this is that the imperative for regulation and self-maintenance in living systems isn't restricted to just one level, such as the integrity of the whole body. Self-maintenance for living systems goes all the way down, even down to the level of individual cells. Every cell in your body—in *any* body—is continually regenerating the conditions necessary for its own integrity over time. The same cannot be said for any current or near-future computer, and would not be true even for a silicon beast machine of the sort I just described.

This shouldn't be taken to imply that individual cells are conscious, or that all living organisms are conscious. The point is that the processes of physiological regulation that underpin consciousness and selfhood in the beast machine theory are bootstrapped from fundamental life processes that apply "all the way down." In this view, it is life, rather than information processing, that breathes the fire into the equations.

EVEN IF ACTUALLY CONSCIOUS machines are far away—if indeed they are possible at all—there is still plenty to worry about. In the nearish future, it is entirely plausible that developments in AI and robotics will deliver new technologies that give the appearance of being conscious, even if there are no conclusive reasons to believe that they actually *are* conscious.

In Alex Garland's 2014 film *Ex Machina,* reclusive billionaire tech genius Nathan invites hotshot programmer Caleb to his remote hideout to meet Ava, the intelligent, inquisitive robot he has created. Caleb's task is to figure out whether Ava is conscious, or whether she—it—is merely an intelligent robot, with no inner life at all.

Ex Machina draws heavily on the Turing test, the famous yardstick for assessing whether a machine can think. In one incisive scene, Nathan is quizzing Caleb about this test. In the standard version of the Turing test, as Caleb knows, a human judge interrogates both a candidate machine and another human, remotely, by exchanging typed messages only. A machine passes the test when the judge consistently fails to distinguish between the human and the machine. But Nathan has something far more interesting in mind. When it comes to Ava, he says, "the challenge is to *show* you that she's a robot—and see if you still feel she has consciousness."

This new game transforms the Turing test from a test of intelligence into a test of consciousness, and as we now know, these are very different phenomena. What's more, Garland shows us that the test is not really about the robot at all. As Nathan puts it, what matters is not whether Ava is a machine. It is not even whether Ava, though a machine, has consciousness. What matters is whether Ava makes a conscious person feel that she (or it) is conscious. The brilliance of this exchange between Nathan and Caleb is that it reveals this kind of test for what it really is: a test of the human, not of the machine. This is true both for Turing's original test and for Garland's twenty-first-century consciousness-oriented equivalent. Garland's dialogue so elegantly captures the challenge of ascribing consciousness to a machine that the term "the Garland test" is now itself gaining traction—a rare example of science fiction feeding back into science.

Many simple computer programs, including chatbots of various kinds, have now been claimed to "pass" the Turing test because a sufficient proportion of human judges have been fooled a sufficient proportion of the time. In one particularly quirky example, also from 2014, ten out of thirty human judges were misled into thinking that a chatbot pretending to be a thirteen-year-old Ukrainian boy was in fact a real thirteen-year-old Ukrainian boy. This led to noisy proclamations that long-standing milestones in AI had finally been surpassed. But of course it is easier to impersonate a foreign teenager who has poor English than it would be to successfully impersonate someone of one's own age, language, and culture, especially when only remote text interactions are allowed. When the chatbot won, its response was "I feel about beating the turing [*sic*] test in quite convenient way." By lowering the bar this far, the test becomes much easier to pass. This was a test of human gullibility, and the humans failed.

As AI continues to improve, the Turing test may soon be passed without such artificially low standards. In May 2020, the research lab OpenAI released GPT-3—a vast artificial neural network trained on examples of natural language drawn from a large swathe of the internet. As well as engaging in chatbot-variety dialogue, GPT-3 can generate substantial passages of text in many different styles when prompted with a few initial words or lines. Although it does not understand what it produces, the fluency and sophistication of GPT-3's output is surprising and, for some, even frightening. In one example, published in the *Guardian*, it delivered a five-hundred-word essay about why humans should not be afraid of AI—ranging across topics from the psychology of human violence to the industrial revolution, and including the disconcerting line: "AI should not waste time trying to understand the viewpoints of people who distrust artificial intelligence for a living."

Despite its sophistication, I am pretty sure that GPT-3 can still be caught out by any reasonably sophisticated human interlocutor. This may not be true for GPT-4, or GPT-10. But even if a future GPT-like system repeatedly aces the Turing test, it would be exhibiting only a very narrow form of (simulated) intelligence—disembodied linguistic exchange—rather than the fully embodied "doing the right thing at the right time" natural intelligence that we see in humans and in many other animals—as well as in my hypothetical silicon beast machine.

When it comes to consciousness, there's no equivalent to the Ukrainian chatbot, let alone to GPT-whatever. The Garland test remains pristine. In fact, attempts to create simulacra of sentient humans have often produced feelings of anxiety and revulsion, rather than the complex mix of attraction, empathy, and pity that Caleb feels for Ava in *Ex Machina*.

*   *   *

THE JAPANESE ROBOTICIST Hiroshi Ishiguro has spent decades
building robots that are as similar as possible to human beings. He
calls them "Geminoids." Ishiguro has created Geminoid copies of
himself (see page 268) and his daughter (then six years old), as well
as a Geminoid Japanese-European TV anchorwoman based on a
blend of about thirty different people. Each Geminoid is constructed from detailed 3D body scans and has pneumatic actuators
able to generate a wide range of facial expressions and gestures.
There is precious little AI as such in these devices—they are all
about human mimicry, with possible applications in, among other
things, remote presence or "telepresence." Ishiguro once used his
Geminoid to deliver a forty-five-minute remote lecture to 150 undergraduate students.

Geminoids are undeniably creepy. They are realistic, but not
quite realistic enough. Think of meeting a Geminoid as the opposite
of meeting a cat. With a cat—or an octopus, for that matter—there
is an immediate sense of the presence of another conscious entity,
even though visual appearances are so different. With a Geminoid,
the striking but imperfect physical similarity accentuates feelings
of disconnection and otherness. In one study from 2009, the
most common feeling experienced by visitors meeting a Geminoid
was fear.

This kind of reaction exemplifies the so-called "uncanny valley,"
a concept originated by another Japanese researcher, Masahiro
Mori, in 1970. Mori proposed that as a robot begins to look humanlike, it will elicit increasingly positive and empathetic reactions
from people (think C3PO in *Star Wars*). But once it passes a certain
point, where it appears strikingly human in some ways but falls

Hiroshi Ishiguro with his Geminoid.
Used with permission from ATR Hiroshi Ishiguro Laboratories

short in others, these reactions will turn rapidly to revulsion and fear—the uncanny valley—recovering only as the resemblance becomes even closer, to the point of indistinguishability. There are many theories about why the uncanny valley exists, but there is little doubt that it does.

Although real-world robots find it hard to escape from the uncanny valley, developments in the virtual world are already clambering back up the slope and out the other side. Recent advances in machine learning using "generative adversarial neural networks"—GANNs for short—can generate photorealistic faces of people who never actually existed (see page 269).* These images are created by

---

* These synthetic faces were generated using thispersondoesnotexist.com.

cleverly mixing features from large databases of actual faces, employing techniques similar to those we used in our hallucination machine (described in chapter 6). When combined with "deepfake" technologies, which can animate these faces to make them say anything, and when what they say is powered by increasingly sophisticated speech recognition and language production software, such as GPT-3, we are all of a sudden living in a world populated by virtual people who are effectively indistinguishable from virtual representations of real people. In this world, we will become accustomed to not being able to tell who is real and who is not.

These people are not real.

Anyone who thinks that these developments will hit a ceiling before a video-enhanced Turing test is convincingly passed is likely to be mistaken. To think this way reveals a resistant case of human exceptionalism, a failure of imagination, or both. It will happen. Two questions remain. The first is whether these new virtual creations will be able to cross into the real world, traversing the uncanny valley in which Ishiguro's Geminoids remain trapped. The second is whether the Garland test will also fall. Will we feel that these new agencies are actually conscious, as well as actually intelligent—even when we know that they are nothing more than lines of computer code?

And if we do feel that way, what will that do to us?

THE RAPID RISE OF AI—whatever mixture of hype and reality it is fueled by—has sparked a resurgent and necessary discussion of ethics. Many ethical concerns have to do with the economic and societal consequences of near-future technologies like self-driving cars and automated factory workers, where significant disruption is inevitable.* There are legitimate worries about delegating decision-making capability to artificial systems, the inner workings of which may be susceptible to all kinds of bias and caprice, and which may remain opaque—not only to those affected, but also to those who designed them. At the extreme end of the spectrum, what horror could be unleashed if an AI system were put in charge of nuclear weapons, or of the internet backbone?

There are also ethical concerns about the psychological and behavioral consequences of AI and machine learning. Privacy invasion by deepfakes, behavior modification by predictive algorithms, and belief distortion in the filter bubbles and echo chambers of social media are just a few of the many forces that pull at the fabric of our societies. By unleashing these forces, we are willingly ceding our identities and autonomy to faceless data corporations in a vast uncontrolled global experiment.

Against this background, ethical discussions about machine

---

* Some of these technologies are not as new as they seem. My colleague Ryota Kanai recently offered that "a horse is basically a self-driving horse."

consciousness can appear indulgent and abstruse. But they aren't. These discussions are necessary, even if the machines in question do not (yet) have consciousness. When the Garland test is passed, we will share our lives with entities that we *feel* have their own subjective inner lives, even though we may know, or believe, that they do not. The psychological and behavioral consequences of this are hard to foresee. One possibility is that we will learn to distinguish how we *feel* from how we should *act*, so that it will seem natural to care for a human but not for a robot even though we feel that both have consciousness. It is not clear what this will do to our individual psychologies.

In the TV series *Westworld*, lifelike robots are developed specifically to be abused, killed, and raped—to serve as outlets for humanity's most depraved behaviors. Could it be possible to torture a robot while feeling that it is conscious and simultaneously knowing that it is not, without one's mind fracturing? With the minds we have now, behavior like this would be top-end sociopathic. Another possibility is that the circle of our moral concern will be distorted by our anthropocentric tendency to experience greater empathy for entities toward which we feel greater similarity. In this scenario we may care more about our next-generation Geminoid twins than we do about other humans, let alone about other animals.

Of course, not all futures need be so dystopian. But as the footrace between progress and hype in AI gathers pace, psychologically informed ethics must play its part too. It is simply not good enough to put new technology out there and wait to see what happens. Above all, the standard AI objective of re-creating and then exceeding human intelligence should not be pursued blindly. As Daniel Dennett has sensibly put it, we are building "intelligent tools, not colleagues," and we must be sure to recognize the difference.

And then comes the possibility of true machine consciousness. Were we to wittingly or unwittingly introduce new forms of subjective experience into the world, we would face an ethical and moral crisis on an unprecedented scale. Once something has conscious status, it also has moral status. We would be obliged to minimize its potential suffering in the same way we are obliged to minimize suffering in living creatures, and we're not doing a particularly good job at that. And for these putative artificially sentient agents, there is the additional challenge that we might have no idea what kinds of consciousness they might be experiencing. Imagine a system subject to an entirely new form of suffering, for which we humans have no equivalent or conception, nor any instincts by which to recognize it. Imagine a system for which the distinction between positive and negative feelings does not even apply, for which there is no corresponding phenomenological dimension. The ethical challenge here is that we would not even know what the relevant ethical issues were.

However far away real artificial consciousness remains, even its remote possibility should be given some consideration. Although we do not know what it would take to create a conscious machine, we also do not know what it would *not* take.

In June 2019, the German philosopher Thomas Metzinger called for an immediate thirty-year moratorium on all research aimed at generating what he called "synthetic phenomenology," precisely for these reasons. I was there when he made his announcement. We were both speaking at a meeting on artificial consciousness hosted by the Leverhulme Centre for the Future of Intelligence, in Cambridge. Metzinger's entreaty is difficult to follow to the letter, since much if not all computational modeling in psychology could fall under his umbrella, but the thrust of his message is clear. We should not

blithely forge ahead attempting to create artificial consciousness simply because we think it's interesting, useful, or cool. The best ethics is
preventative ethics.

In the heyday of vitalism, it might have seemed as preposterous
to talk about the ethics of artificial life as the ethics of artificial
consciousness can seem to us today. But here we are, a little over a
hundred years later, with not only a deep understanding of what
makes life possible, but many new tools to modify and even create
it. We have gene-editing techniques like CRISPR, which enables
scientists to easily alter DNA sequences and change the function of
genes. We even have the capability to develop fully synthetic organisms built from the "genes up": in 2019, researchers in Cambridge
created a variant of *Escherichia coli* with a fully synthetic genome.
The ethics of creating new forms of life is suddenly very relevant
indeed.

And perhaps it will be biotechnology, rather than AI, that
brings us closest to synthetic consciousness. Here, the advent of
"cerebral organoids" is of particular significance. These are tiny
brain-like structures, made of real neurons, which are grown from
human pluripotent stem cells (cells which can differentiate into
many different forms). Although not "mini brains," cerebral organoids resemble the developing human brain in ways which make
them useful as laboratory models of medical conditions in which
brain development goes wrong. Could these organoids harbor a
primitive form of bodiless awareness? It is hard to rule the possibility out, especially when they start to show coordinated waves of
electrical activity not unlike those seen in premature human babies,
as one recent study found.

Unlike computers, cerebral organoids are made out of the same
physical stuff as real brains, removing one obstacle to thinking of

them as potentially conscious. On the other hand, they remain extremely simple, they are completely disembodied, and they do not interact with the outside world at all (though it is possible to wire them up to cameras and robotic arms and the like). For my money, while current organoids are highly unlikely to be conscious, the question will remain disconcertingly open as the technology develops. This brings us back to a need for preventative ethics. The possibility of organoid consciousness has ethical urgency not only because it cannot be ruled out, but because of the potential scale involved. As the organoid researcher Alysson Muotri has said, "We want to make farms of these organoids."

WHY IS THE PROSPECT OF MACHINE consciousness so alluring? Why does it exert such a pull on our collective imagination? I've come to think that it has to do with a kind of techno-rapture, a deep-seated desire to transcend our circumscribed and messily material biological existence as the end times approach. If conscious machines are possible, with them arises the possibility of rehousing our wetware-based conscious minds within the pristine circuitry of a future supercomputer that does not age and never dies. This is the territory of mind uploading, a favorite trope of futurists and transhumanists for whom one life is not enough.

Some even think we may already be there. The Oxford University philosopher Nick Bostrom's "simulation argument" outlines a statistical case proposing that we are more likely to be part of a highly sophisticated computer simulation, designed and implemented by our technologically superior and genealogically obsessed descendants, than we are to be part of the original biological human race. In this view, we already are virtual sentient agents in a virtual universe.

Some captivated by the techno-rapture see a fast-approaching Singularity, a critical point in history at which AI is poised to bootstrap itself beyond our understanding and outside our control. In a post-Singularity world, conscious machines and ancestor simulations abound. We carbon-based life-forms will be left far behind, our moment in the sun over and done.

It doesn't take much sociological insight to see the appeal of this heady brew to our technological elite who, by these lights, can see themselves as pivotal in this unprecedented transition in human history, with immortality the prize. This is what happens when human exceptionalism goes properly off the rails. Seen this way, the fuss about machine consciousness is symptomatic of an increasing alienation from our biological nature and from our evolutionary heritage.

The beast machine perspective differs from this narrative in almost every way. In my theory, as we've seen, the entirety of human experience and mental life arises because of, and not in spite of, our nature as self-sustaining biological organisms that care about their own persistence. This view of consciousness and human nature does not exclude the possibility of conscious machines, but it does undercut the amped-up techno-rapture narrative of soon-to-be-sentient computers that propels our fears and permeates our dreams. From the beast machine perspective, the quest to understand consciousness places us increasingly within nature, not further apart from it.

Just as it should.

EPILOGUE

IN JANUARY 2019, I came face-to-face with a living human brain for the first time. Twenty-odd years after I'd first started to research the science of consciousness, ten years since our laboratory at Sussex had opened its doors, and three years after my own anesthesia-induced oblivion with which this book began. After all this time, gazing at the gently pulsing gray-white cortical surface, delicately threaded with dark red veins, it seemed again inconceivable that such a lump of stuff could give rise to an inner universe of thoughts, feelings, perceptions—to a life lived fully in the first person. The profound sense of wonder I felt mixed uncomfortably in my mind with the old joke that a brain transplant is the only operation for which it's better to be the donor than the recipient.

I was the guest of Michael Carter, a pediatric brain surgeon working at the Bristol Royal Hospital for Children, in the west of England. He'd invited me to observe one of the more dramatic neurosurgical procedures carried out anywhere. The patient, a child of just over six, was scheduled for a hemispherotomy. He'd been suffering from severe epilepsy since he was born. The seizures originated from his right cortical hemisphere, which had been badly damaged during his premature birth. All standard antiseizure medications had failed, so as a last resort the neurosurgeon was called in.

A hemispherotomy involves complete neural disconnection of the brain's dysfunctional right hemisphere. The surgeon enters the brain through the right-hand side, removes ("resects") the temporal lobe, and then cuts through all the bundles of connections—the white-matter tracts—that link the right hemisphere with the rest of the brain and body. The isolated hemisphere remains inside the skull, and is still connected to its blood supply. It is a living but isolated island of cortex. An extreme version of the more familiar split-brain operation, the idea is that complete neural disconnection prevents electrical storms originating in the damaged right hemisphere from spreading to the rest of the brain. If the operation is carried out early enough, the young brain is often sufficiently adaptable that the remaining hemisphere can pick up most or all of the slack. Despite the radical nature of this surgery, and although every case is different, outcomes are generally good.

This particular operation started at about noon and lasted for a shade over eight hours. In my own work I can barely last five minutes before being distracted by an email or the cricket score, or by making another cup of tea. Michael worked without pause for hour after hour, patiently, methodically, unrelentingly, supported by a neurosurgical trainee and a rotating team of assistants. About half-way through, when the trainee surgeon took a short break, I was invited to scrub in and step up to the surgical microscope. I had not expected such a privileged view. Peering into the brightly lit cavities of the child's brain, I tried to register my abstract knowledge of different regions and pathways to the welter of tissue illuminated before me. It made little sense. The crisp cortical hierarchies and counterflowing weaves of bottom-up and top-down signaling that I knew from my research were nowhere to be seen. The brain was newly inscrutable, and I was left in awe both of the neurosurgeon's skill and of the material reality of this most magical of objects. It felt almost transgressive. A curtain had been pulled back revealing something too intimate to be so openly on view. I was looking directly into the mechanics of a human self.

THE SURGERY WENT ACCORDING TO PLAN. Some time after eight o'clock, Michael left the trainee to finish stitching the scalp back together and took me along to meet the child's family. They were grateful and relieved. I wondered what they would have been feeling had they seen what I'd seen that day.

Later, driving home through the winter darkness, my thoughts returned to David Chalmers's description of the hard problem of consciousness: "It is widely agreed that experience arises from a physical basis, but we have no good explanation of why and how it so arises. Why should physical processing give rise to such a rich inner life at all? It seems objectively unreasonable that it should, and yet it does."

Faced with this mystery, philosophy has provided a range of options, from panpsychism (consciousness everywhere, more or less), to eliminative materialism (no consciousness, at least not how we think of it), and everything in between. But the science of consciousness isn't about choosing from a set menu, however swanky the restaurant or skilled the chef. It's more like cooking with whatever you can find in the fridge, where various bits and pieces from philosophy, neuroscience, psychology, computer science, psychiatry, machine learning, and so on are combined and recombined in different ways, and turned into something new.

This is the essence of the real problem approach to consciousness. Accept that consciousness exists, and then ask how the various phenomenological properties of consciousness—which is to say how conscious experiences are structured, what form they take—relate to properties of brains, brains that are embodied in bodies and embedded in worlds. Answers to these questions can begin by identifying correlations between this-or-that pattern of brain activity and this-or-that type of conscious experience, but they need not and should not end there. The challenge is to build increasingly sturdy explanatory bridges between mechanism and phenomenology, so that the relations we draw are not arbitrary but make sense. What does "make sense" mean in this context? Again: explain, predict, and control.

Historically, this strategy echoes how our scientific understanding of life transcended the magical thinking of vitalism by individuating the properties of living systems, and then accounting for each in terms of their underlying mechanisms. Life and consciousness are of course different things, though I hope by now I've persuaded you that they are more intimately connected than at first they might seem. Either way, the strategy is the same. Instead of attempting to solve the hard problem of consciousness head-on, and rather than sidelining the experiential qualities of consciousness altogether, the real problem approach offers genuine hope of reconciling the physical with the phenomenal—dissolving, not solving the hard problem.

We began with conscious *level*—the difference between being in a coma and being wide awake and aware—where we focused on the importance of *measurement*. The key point here is that candidate measures, like causal density and integrated information, are not arbitrary. Rather, they capture highly conserved properties of *all* conscious experiences, namely that every conscious experience is simultaneously unified and distinct from all other conscious experiences. Every conscious scene is experienced "all of a piece," and every experience is the way it is, and not some other way.

We then moved on to the nature of conscious *content* and in particular the experience of being a conscious *self*. I posed a series of challenges to the way things seem that in each case encouraged us to adopt new, post-Copernican perspectives on conscious perception.

The first challenge was to understand perception as an active, action-oriented construction, rather than as a passive registration of an objective external reality. Our perceived worlds are both less than and more than whatever this objective external reality might be. Our brains create our worlds through processes of Bayesian best guessing in which sensory signals serve primarily to rein in our continually evolving perceptual hypotheses. We live within a controlled hallucination which evolution has designed not for accuracy but for utility.

The second challenge turned this insight inward, to the experience of being a self. We explored how the self is itself a perception, another variety of controlled hallucination. From experiences of personal identity and continuity over time, all the way down to the inchoate sense of simply being a living body, these pieces-of-selfhood all depend on the same delicate dance between inside-out perceptual prediction and outside-in prediction error, though now much of this dance takes place within the confines of the body.

The final challenge was to see that the predictive machinery of conscious perception has its origin and primary function not in representing the world or the body, but in the control and regulation of our physiological condition. The totality of our perceptions and cognitions—the whole panorama of human experience and mental life—is sculpted by a deep-seated biological drive to stay alive. We perceive the world around us, and ourselves within it, *with*, *through*, and *because of* our living bodies.

This is my theory of the beast machine, a twenty-first-century version—or inversion—of Julien Offray de La Mettrie's "*l'homme machine*." And it's here that the deepest shifts in how to think about consciousness and selfhood take place.

There is the puzzle that experiences of "being a self" are very different from experiences of the world around us. Now we can understand them as different expressions of the same principles of perceptual prediction, with the differences in phenomenology tracing back to differences in the kinds of prediction that are involved. Some perceptual inferences are geared toward finding out about objects in the world, while others are all about controlling the interior of the body.

By tying our mental lives to our physiological reality, age-old conceptions of a continuity between life and mind are given new substance, buttressed by the sturdy pillars of predictive processing and the free energy principle. And this deep continuity in turn allows us to see ourselves in closer relation to other animals and to the rest of nature, and correspondingly distant from the fleshless calculus of AI. As consciousness and life come together, consciousness and intelligence are teased apart. This reorientation of our place in nature applies not only to our physical, biological bodies, but to our conscious minds, to our experiences of the world around us and of being who we are.

* * *

Every time science has displaced us from the center of things, it has given back far more in return. The Copernican revolution gave us a universe—one which astronomical discoveries of the last hundred years have expanded far beyond the limits of human imagination. Charles Darwin's theory of evolution by natural selection gave us a family, a connection to all other living species and an appreciation of deep time and of the power of evolutionary design. And now the science of consciousness, of which the beast machine theory is just one part, is breaching the last remaining bastion of human exceptionalism—the presumed specialness of our conscious minds—and showing this, too, to be deeply inscribed into the wider patterns of nature.

Everything in conscious experience is a perception of sorts, and every perception is a kind of controlled—or controlling— hallucination. What excites me most about this way of thinking is how far it may take us. Experiences of free will are perceptions. The flow of time is a perception. Perhaps even the three-dimensional structure of our experienced world and the sense that the contents of perceptual experience are objectively real—these may be aspects of perception too. The tools of consciousness science are allowing us to get ever closer to Kant's noumenon, the ultimately unknowable reality of which we, too, are a part. All these ideas are testable, and whichever way the data come out, simply posing questions of this kind reshapes our understanding of what consciousness is, how it happens, and what it is for. Every step chips away at the beguiling but unhelpful intuition that consciousness is one thing—one big scary mystery in search of one big scary solution.

There are plenty of practical implications too. Theoretically inspired measures of conscious level are ushering in new conscious-

ness "meters" that are increasingly able to detect residual awareness—
"covert consciousness"—in behaviorally unresponsive patients. Computational models of predictive perception are shedding new light on
the basis of hallucinations and delusions, inaugurating a transformation in psychiatry from treating symptoms to addressing causes. And
there are all sorts of new directions for AI, brain-machine interfaces
and virtual reality, among an abundance of established and emerging
technologies. Going after the biological basis of consciousness is a
surprisingly useful thing to do.

All this being said, facing up to the mystery of awareness is, and
always will be, a deeply personal journey. What good is a science of
consciousness unless it sheds new light on our individual mental
lives, and on the inner lives of those around us?

This is the real promise of the real problem. Wherever it eventually takes us, following this road will lead us to understand so
many new things about conscious experiences of the world around
us, and of ourselves within it. We will see how our inner universe is
part of, and not apart from, the rest of nature. And, though we may
not think of it as often as we might, we will have the chance to
make a new peace with what happens—or does not happen—when
the controlled hallucination of *being you* finally breaks down into
nothingness. When oblivion is not an anesthesia-induced interrup-
tion to the river of consciousness, but a return to the eternity that
each of us at one time emerged from.

At the end of *this* story, when life in the first person reaches its
conclusion, perhaps it's not so bad if a little mystery remains.

ACKNOWLEDGMENTS

The ideas in this book have been shaped thanks to countless conversations with friends, colleagues, students, teachers, and mentors over more than two decades. I thank them all.

Thank you to all the past and present members of my research group at Sussex. It's been, and continues to be, a privilege to work with you. I am particularly grateful to those from the group whose research I have drawn on in this book. Thank you, Lionel Barnett, Adam Barrett, Peter Lush, Alberto Mariola, Yair Pinto, Warrick Roseboom, Michael Schartner, David Schwartzman, Maxine Sherman, Keisuke Suzuki, and Alexander Tschantz.

Thank you also, Manuel Baltieri, Reny Baykova, Luc Berthouze, Daniel Bor, Chris Buckley, Acer Chang, Paul Chorley, Ron Chrisley, Andy Clark, Marianne Cole, Clémence Compain, Guillaume Corlouer, Hugo Critchley, Zoltán Dienes, Tom Froese, Paul Graham, Inman Harvey, Owen Holland, Ryota Kanai, Tomasz Korbak, Isabel Maranhão, Federico Micheli, Beren Millidge, Thomas Nowotny, Andy Philippides, Charlotte Rae, Colin Reveley, Ryan Scott, Lina Skora, Nadine Spychala, Marta Suarez-Pinilla, Chris Thornton, Hao-Ting Wang, and Jamie Ward—all colleagues and friends, Sussex past and present, whose work and ideas have contributed much to my own, and many of whom have been kind enough to read and comment on parts of this book.

I'm very grateful to have been mentored by outstanding scientists at every stage of my career. Nicholas Mackintosh, who died in 2015, saw me through my undergraduate degree at Cambridge and gave me the confidence to set out in academia. Phil Husbands supervised my DPhil. (what they used to call a PhD at Sussex), allowing me the freedom to explore while keeping a careful eye on where I was going. Gerald Edelman, who died in 2014, mentored me for more than six years as a postdoctoral researcher. Under his guidance, my interest in consciousness finally became my work. I would also like to thank Margaret Boden, Andy Clark, Daniel Dennett, and Thomas Metzinger, who have all been extraordinary influences and inspirations over many years.

Several people took on the task of reading and commenting in detail on early full drafts of this book. Heartfelt thanks to Tim Bayne, Andy Clark (again), Claudia Fischer, Jakob Hohwy, and Murray Shanahan. Their ideas and suggestions have benefited me enormously. I am also extremely grateful to Karl Friston, Marcello Massimini, Thomas Metzinger (again), Adrian Owen, and Aniruddh (Ani) Patel, who gave detailed advice on selected parts. Many thanks to Steve West (Lazy Chief) for his stylish illustrations. Thank you to Baba Brinkman, whose gift with words is a continuing inspiration. Working with him on his *Rap Guide to Consciousness* has been a real highlight. Thank you to Michael Carter for bringing me into his operating theater to witness a hemispherotomy. Thank you, David Edelman and Graziano Fiorito, for bringing me into the world of octopuses; Ian Cheney and Laurie Santos, for bringing me into the world of monkeys; and Luise Schreiter, for bringing me into another world altogether.

Of the many others whose advice, ideas, and support have over many years helped me, I can mention only a few. Thank you, Anil Ananthaswamy, Chris Anderson, Bernard Baars, Lisa Feldman Barrett, Isabel Behncke, Tristan Bekinschtein, Yoshua Bengio, Matt Bergman, Heather Berlin, David Biello, Robin Carhart-Harris, Olivia Carter, David Chalmers, Craig Chapman, Axel Cleeremans, Athena Demertzi, Steve Fleming, Zaferios Fountas, Friday Football, Chris Frith, Uta Frith, Alex Garland, Mariana Garza, Mel Goodale, Annaka Harris, Sam Harris, Nick Humphrey, Rob Illife, Robin Ince, John Iversen, Eugene Izhikevich, Alexis Johansen, Robert Kentridge, Christof Koch, Sid Kouider, Jeff Krichmar, Victor Lamme, Hakwan Lau, Steven Laureys, Rafael Malach, Daanish Masood, Simon McGregor, Pedro Mediano, Lucia Melloni, Liad Mudrick, Phil Newman, Angus Nisbet, the Parasites, Megan Peters, Giovanni Pezzulo, Tony Prescott, Blake Richards, Fernando Rosas, Adam Rutherford, Tim Satterthwaite, Tom Smith, Narayanan Srinivasan, Catherine Tallon-Baudry, Giulio Tononi, Nao Tsuchiya, Nick Turk-Browne, Lucina Uddin, Simon van Gaal, Bruno van Swinderen, Anniek Verholt, Paul Verschure, Lucy Walker, Nigel Warburton, Lisa Westbury, and Martijn Wokke.

These people have set me straight in all sorts of ways over the years. This is not to say that they agree with what I say. Any remaining errors and idiocies in the book are of course entirely my own.

I could not have written this book without support from the Wellcome Trust (via an Engagement Fellowship), the Dr. Mortimer and Theresa Sackler Foundation, and the Canadian Institute for Advanced Research (for whom I co-direct their program on

Brain, Mind, and Consciousness). I am grateful for their support both for my time and for the research my group carries out. And thank you to Sussex University for providing me with an academic home for many years now.

My final thanks are reserved for my agent and my editors. Will Francis, of Janklow & Nesbit, encouraged me to write this book in the first place, helped me shape the proposal, find excellent editors, and generally shepherded the whole project from start to finish. Equal thanks go to those excellent editors, Rowan Cope and Laura Hassan at Faber & Faber, Stephen Morrow at Dutton/Penguin, and copy editor Eleanor Rees.

NOTES

PROLOGUE

9 meditation on mortality: Barnes (2008).

1. THE REAL PROBLEM

15 "global workspace" theory: Baars (1988); Dehaene & Changeux (2011); Mashour et al. (2020); Shanahan (2010).

15 degree of behavioral flexibility: Another way to think of global workspace theory is as a theory of "access consciousness," as compared to "phenomenal consciousness." Phenomenal consciousness refers explicitly to experience. Access consciousness emphasizes cognitive functions over experience. When a mental state is "access conscious," it means that the mental state is available for all sorts of cognitive functions including reasoning, decision-making, and the control of behavior. See Block (2005).

15 "higher-order thought" theory: There are many varieties of higher-order theory. Some address phenomenology more directly than others. See R. Brown et al. (2019); Fleming (2020); Lau & Rosenthal (2011).

16 definition of a "gene" has changed: Portin (2009).

17 Here is how he describes it: Chalmers (1995a), p. 201.

17 Chalmers contrasts this: More recently, Chalmers has introduced the "meta-problem" of consciousness, which is the problem of why people think there is a hard problem of consciousness (Chalmers, 2018). The meta-problem is actually a version of the easy problem, since it is about explaining behavior—in this case the verbal behavior of people who express belief in a hard problem of consciousness. One thing I like about the meta-problem is that one can appreciate that it is a problem, and study it, whatever metaphysical stance one takes about consciousness itself.

18 A "mechanism": Craver & Tabery (2017).

18 "[E]ven when we have explained": Chalmers (1995a), p. 203, italics in original.

19 they can be treated synonymously: The difference between physicalism and materialism is largely historical. "Materialism" is an older word than "physicalism." Some people also argue that physicalism is a "linguistic thesis" (that every linguistic statement is equivalent to some physical statement), whereas materialism is a more general claim about the nature of things. See Stoljar (2017).

19 **take physicalism for granted:** Although most functionalists are physicalists, it is possible to be a functionalist without being a physicalist.

20 **Brains are very different:** Matthew Cobb's engrossing *The Idea of the Brain* (2020) relates the history of how brain function has often been interpreted using the dominant technology of the time (and sometimes the other way around).

20 **A computer that plays Go:** Silver et al. (2017). The story of the original program, AlphaGo, is beautifully told in a film of the same name: https://www.alphagomovie.com/. Some might quibble that these programs are more accurately described as playing "the history of Go" rather than Go itself.

21 **there's a valid question:** A more sophisticated version of this argument has been developed by John Searle in his famous "Chinese room" thought experiment. I didn't use this example here because Searle's argument is targeted primarily at intelligence (or "understanding") rather than consciousness (Searle, 1980).

21 **an empirical dead end:** The philosopher John Perry said: "If you think about consciousness long enough, you either become a panpsychist or you go into administration." Perhaps the problem lies in only thinking, rather than doing (science). For an articulate defense of panpsychism, I recommend *Galileo's Error* by Philip Goff (2019). Perry's quote appears in a 2018 article in *Quartz* by Olivia Goldhill: qz.com/1184574/the -idea-that-everything-from-spoons-to-stones-are-conscious-is-gaining-academic -credibility.

21 *mysterianism*: McGinn (1989).

22 **no human could ever understand:** For an argument against the view that some things will always lie beyond human understanding, see Deutsch (2012).

23 **most intellectually honest:** There is one more "ism" which deserves mention, though it is a more informal "ism" than those discussed in the main text. *Illusionism* is the view that (phenomenal) consciousness is an introspective illusion—that when we introspect about conscious states, we misrepresent them as having phenomenal properties—qualia— that in fact they do not have. In one reading, which I disagree with, illusionism says that conscious states do not really exist. In another, which I am more sympathetic to, illusionism says that conscious experiences exist but are not what we think they are. It is possible, though not clear, that this reading of illusionism is compatible with what I will say in this book. For more on illusionism, see Frankish (2017), and see Graziano (2017) for a theory in this spirit.

23 **any conscious experience whatsoever:** There are at least two types of philosophical zombie. "Behavioral zombies" are indistinguishable from their conscious counterparts from the outside—in terms of their behavior. "Neurological zombies" add the additional fantasy of being identical to a conscious creature also from the inside. A neurological zombie has the same internal structure, and may also be made out of the same electrochemical wetware as the conscious creature that it is a zombie version of. All varieties of zombie entirely lack consciousness.

24 **a vast network:** Gidon et al. (2020); Herculano-Houzel (2009).

25 **imagine the unimaginable:** Zombie defenders may respond that it is only their logical possibility that matters, not their conceivability given the laws of physics in this particular universe. I disagree. A backward-flying A380 may be logically possible given alternative principles of aerodynamics, but accepting this doesn't shed any light on how a real A380 actually flies, in this world, with the laws of physics and aerodynamics that we actually

have. What I want to know is how a real brain (and body, and so on), abiding by the laws of physics in *this* universe, actually shapes and gives rise to conscious experience in *this* universe.

25 **the insights of many others:** The "real problem" way of putting things is not new—at least not entirely. Chalmers himself described similar strategies in the form of the "mapping problem" (Chalmers, 1996) and the "principle of structural coherence" (Chalmers, 1995a)—which forms part of his original description of the hard problem. There is also a long and highly influential body of work in "neurophenomenology" which attempts to match phenomenological properties to aspects of the brain and its activity (Thompson, 2014; Varela et al., 1996). There are, however, differences in emphasis between these positions (Seth, 2009, 2016b).

28 **Nothing worth reading has been written:** Sutherland (1989).

28 **become the dominant method:** Crick & Koch (1990). Around the same time, the American philosopher Daniel C. Dennett published his influential *Consciousness Explained* (1991). Reading this book in the early 1990s was, for me, a turning point. It remains a rambunctious and enlightening read. See LeDoux, Michel, & Lau (2020) and Seth (2017; 2018) for more on the history of consciousness science.

28 **"minimal neuronal mechanisms":** Crick & Koch (1990).

29 **NCC for that particular experience:** Note that NCCs are usually interpreted as relating to specific brain regions, but this need not be the case. The definition of an NCC relates to a neural mechanism which could be implemented across a range of brain regions. For a given NCC, the brain circuits involved may even change over time (G. M. Edelman & Gally, 2001).

30 **The brain activity that goes along:** A subtlety here is that brain regions that show up in binocular rivalry studies are often associated with *transitions* between conscious perceptions, which is important because perception of change is not the same thing as change of perception. See Blake et al. (2014). This distinction is revisited in chapter 6.

30 **that it is difficult:** These problems have been talked about for years, but were generally either dismissed as irrelevant or otherwise swept under the carpet. Two papers from 2012 finally crystallized the issue: Aru et al. (2012); de Graaf et al. (2012).

30 **should not be confused:** Some experiments have made valiant attempts to distinguish conscious perception from attention, and from behavioral report. The results from so-called "no report" paradigms, in which volunteers do not make behavioral reports about what they perceive, are particularly intriguing. In many of these studies, the remaining NCC does not include the frontal brain regions that are central to theories like the global workspace theory and higher-order thought theory. See Frässle et al. (2014); Tsuchiya et al. (2015), and Raccah et al. (2021) for a recent discussion.

31 **fully satisfying science:** Another way of putting this ambition comes from the philosophers Susan Hurley and Alva Noë, who distinguish between "comparative" explanatory gaps, which have to do with explaining why different experiences have the specific phenomenological properties that they do, and the "absolute" explanatory gap, which is the (hard) problem of why and how there is such a thing as phenomenology at all. We can think of the real problem as exhaustively addressing comparative explanatory gaps in order to resolve, and perhaps dissolve, the absolute explanatory gap. See Hurley & Noë (2003).

31 **Vitalists thought:** Vitalism holds that "living organisms are fundamentally different from non-living entities because they contain some non-physical element or are governed by

different principles than are inanimate things" (Bechtel & Williamson, 1998). Even today a majority of preschool-age children tend to prefer vitalist explanations of life over other more modern explanations (Inagaki & Hatano, 2004). The historical parallels between vitalism and the science of consciousness have been particularly vigorously explored by the philosopher Patricia Churchland (Churchland, 1996).

33 **one humdinger eureka:** The yearning for a eureka solution may partly account for the persistent appeal of theories of consciousness based on quantum mechanics, most of which trace back to the mathematician Roger Penrose's *The Emperor's New Mind*, published in 1989. While it can't be ruled out that some future quantum-based theory may have something useful to say about consciousness, the attempts so far seem to me to evince a false syllogism: Quantum mechanics is mysterious, consciousness is mysterious, therefore they must be related.

2. MEASURING CONSCIOUSNESS

36 **Joachim Dalencé:** www.encyclopedia.com/science/dictionaries-thesauruses-pictures-and-press-releases/dalence-joachim.

37 *Inventing Temperature:* H. Chang (2004).

37 **Could the same approach work:** Possibly the first person to ask, and answer, this question was Thomas Henry Huxley ("Darwin's Bulldog"). In an 1870 lecture he said, "I believe that we shall, sooner or later, arrive at a mechanical equivalent of consciousness, just as we have arrived at a mechanical equivalent of heat." Quoted in Cobb (2020), p. 113.

37 **potential to transform scientific understanding:** Seth et al. (2008).

38 **more than four million:** Weiser et al. (2008).

38 **"bispectral index" monitor:** Myles et al. (2004).

39 **monitors have remained controversial:** Nasraway et al. (2002).

40 **four times as many neurons:** Herculano-Houzel (2009).

40 **is no reason to doubt:** See Lemon & Edgley (2010). The cerebellum often shows up as an "area of interest" in neuroimaging studies. Often these findings are not discussed because the cerebellum wasn't the focus of the experiment, so researchers don't know what to say about it.

40 **differences in the brain's energy consumption:** DiNuzzo & Nedergaard (2017).

42 **complexity of these patterns:** Ferrarelli et al. (2010); Massimini et al. (2005).

43 **a compressed representation:** To create a compressed representation of a sequence of all 1s (or all 0s), for a sequence of length n, all you need to do is specify the initial 1 (or 0) and then specify n repetitions. At the other extreme, an entirely random sequence cannot be compressed at all: you have to specify every single 1 and 0 in its proper place in order to accurately regenerate it. Sequences that contain some amount of predictable structure will have compressed representations somewhere in the middle. The length of the optimally compressed representation is called the Kolmogorov-Chaitin-Solomonoff complexity. LZW complexity approximates and provides an upper bound on this quantity.

43 **landmark study:** Casali et al. (2013).

44 **series of studies:** Schartner et al. (2017b); Schartner et al. (2015). We made sure, in these studies, that changes in LZW complexity were not trivial reflections of other—better known—changes in EEG brain signals that accompany loss of consciousness, like increases in the power of low-frequency "delta" waves, which is known to happen during sleep.

44 **same pattern of results:** Casarotto et al. (2016).

45 **clinical approaches:** See for example the widely used Glasgow coma scale (Teasdale & Murray, 2000).

45 **This rare affliction:** Locked-in syndrome is most often caused by damage to the pons—a part of the brainstem which contains important neuronal pathways linking the spinal cord with other parts of the brain.

45 **wrote an entire book:** Bauby (1997). You might think that being locked-in is a fate worse than death, but while some people in this condition do find life intolerable, a surprising proportion report a reasonable quality of life (Rousseau et al., 2015). This underscores the dangers of assuming how things are on the inside from how they appear on the outside.

47 **a strange case:** The story is also related here: www.humanbrainproject.eu/en/follow-hbp /news/measures-of-consciousness-and-a-tale-of-cultural-loss.

47 **Owen concluded:** Owen et al. (2006). Adrian Owen recounts the story of his team's discovery, and its implications for medicine and consciousness research, in his book *Into the Gray Zone* (Owen, 2017).

48 **laborious way of communicating:** Monti et al. (2010). Given that patients can answer only "yes" or "no" in this setup, researchers have to be very careful when selecting their questions. Asking whether someone is in pain is reasonable, but what about asking whether they want to remain alive? There are many ethical quandaries hereabouts. In 2016, the writer Linda Marshall Griffiths, the producer Nadia Molinari, and I created an audio drama, *The Sky Is Wider,* which explores these issues. https://www.lindamarshall griffiths.co.uk/the-sky-is-wider-best-single-drama.

48 **a recent analysis:** Naci et al. (2017).

49 **may be limited:** A better-known example of this problem has been the attempt to define and measure intelligence with respect to a single scale, such as "IQ." This project has always floundered in the face of the sheer variety of intelligent behavior seen across individuals, cultures, and species. See Bayne, Hohwy, and Owen (2016) for a multidimensional approach to characterizing conscious level.

50 **degree of reflective insight:** See Konkoly et al. (2021).

50 **communicate with people during lucid dreams:** See Konkoly et al. (2021).

51 **" . . . Little by little":** Albert Hofmann, "LSD—My Problem Child," http://psychedelic -library.org/child1.htm.

52 **a lost generation:** See Michael Pollan's *How to Change Your Mind* for much more on this history (Pollan, 2018).

52 **striking alterations in brain dynamics:** Carhart-Harris et al. (2012).

53 **high time resolution:** MEG and EEG data typically have very high time resolution, on the order of milliseconds, since they directly reflect the electrical activity of populations of neurons. Functional MRI is very slow by comparison, with a natural timescale of seconds. This is partly because of the slow operation of a typical MRI scanner, typically allowing one measurement every one or two seconds, and partly because of the slow timescale of the blood oxygenation signal that fMRI measures.

53 **results were clear:** Schartner et al. (2017a). Another, more recent study found the same pattern of results with DMT (Timmermann et al., 2019).

54 **Our findings:** In a follow-up study, led by Lionel Barnett, we found that the psychedelic state involves significant reductions in "information flow" between regions of the cortex. This again speaks to a loss of perceptual structure in the psychedelic state. See Barnett et al. (2020).

55 A paper published in 1998: Tononi & Edelman (1998). Edelman died in 2014—I wrote one of many obituaries paying tribute to this hugely influential scientist (Seth, 2014a).

57 seem tied together: Some philosophers have questioned the assumption that conscious experiences are necessarily unified (Bayne, 2010). One scenario where the possibility of non-unified consciousness is often raised is in split-brain patients, who have had their cortical hemispheres surgically separated. More on this in chapter 3.

60 "neural complexity": Tononi et al. (1994).

60 "causal density": Seth et al. (2011a); Seth et al. (2006).

60 refining the mathematics: For a comparison of measures of complexity in the context of neuroscience, see Seth et al. (2011a) and Mediano et al. (2019).

61 Other measures: Demertzi et al. (2019); Luppi et al. (2019).

3. PHI

62 IIT says that subjective: Some of the main papers on IIT from Tononi's group are: Tononi (2008); Tononi (2012); Tononi et al. (2016). For an accessible manifesto, see Koch (2019).

62 recent paper of mine: Seth et al. (2006).

63 "a gigantic step": www.scientificamerican.com/article/is-consciousness-universal.

63 dangerous intuition: See for example Scott Aaronson's critiques: www.scottaaronson.com/blog/?p=1799.

64 main claim of the theory: Φ can also be thought of as a way to measure the property of "emergence." Emergence is a very general concept referring to how macroscopic properties (such as a flock) arise from or relate to their microscopic components (the individual birds). See Hoel et al. (2013); Rosas et al. (2020); Seth (2010).

64 ultimate expression of a temperature-based view: Tononi and Koch use this analogy themselves (Tononi & Koch, 2015).

65 ways a system can fail: These are adapted from Tononi (2008).

66 work just as well: This is to a first approximation, setting aside things like contrast adjustment which might operate across the whole array.

66 surgery to divide: Split-brain operations involve severing the massive bundle of nerve fibers that interconnect the cortical hemispheres, the corpus callosum. The surgery can be successful in alleviating severe epilepsy, but is rarely performed now that other, less invasive treatment options are increasingly available. Split-brain patients retain some degree of connectivity between the hemispheres, but for the purposes of this example, let's imagine their entire brain is cut in half. See de Haan et al. (2020).

67 IIT accounts neatly: Tononi et al. (2016).

67 IIT proposes axioms: Besides the axioms of integration and information, IIT proposes three others: that consciousness exists, that it is composed of many elements, and that it is exclusive to a particular spatiotemporal scale (Tononi et al., 2016). The philosopher Tim Bayne has critiqued IIT on the basis that its proposed axioms, particularly the final one regarding "exclusion," may not in fact be self-evidently true (Bayne, 2018).

68 measure the reduction in uncertainty: In information theory, information—reduction of uncertainty—is measured using a quantity called entropy. Entropy (usually termed S) is a function of the number of distinct states a system can be in, together with the probability of being in each of these states. It is given by the equation $S = -\sum p_k \log(p_k)$. In words, this means that for each state (k) of a system, you multiply the probability of being in that state by the logarithm of that probability, then sum these values over all states. For any given system, entropy is highest when each state is equally probable. A fair die has an entropy of about 2.5 bits (when the logarithm is taken in base 2). A loaded die will have a lower entropy.

70 **know everything about a mechanism:** The value of Φ a system has is therefore more a claim about its mechanism (how it is wired up) than about its dynamics (what it does). Indeed, recent versions of IIT describe Φ in terms of "irreducible cause-effect power," which is a claim about mechanism, not dynamics (Tononi et al., 2016).

70 **dividing up the system:** Technically, this division is called the "minimum information partition." There are also tricky issues about how best to deal with partitions of different sizes, since a larger partition will, by virtue of having more elements, be capable of generating more information.

71 **Could an entire country:** According to IIT a "conscious country" might occur if Φ is maximized on a spatiotemporal scale that operates across an entire country, perhaps with individual people being analogous to neurons in a brain. Such a situation would have the peculiar implication that, once a country is conscious, its individual elements—the people—would no longer be individually conscious. This bizarre scenario has been explored by the American philosopher Eric Schwitzgebel. See schwitzsplinters.blogspot.com/2012/03/why-tononi-should-think-that-united.html.

72 **shine light into the brains:** Deisseroth (2015).

73 **actually getting it done:** See www.templetonworldcharity.org/accelerating-research-consciousness-our-structured-adversarial-collaboration-projects. The "inactive" versus "inactivated" experiment was proposed by Umberto Olcese and Giulio Tononi.

73 **"it from bit" view:** Wheeler (1989).

74 **our various versions:** Barrett & Seth (2011); Mediano et al. (2019). Mathematically, our measures work with the empirical distribution of a system, rather than with its maximum entropy distribution.

75 **how it develops:** One intriguing proposal is that the pervasive "spatiality" of visual experience is explained by the grid-like anatomy found in lower levels of the visual cortex (Haun & Tononi, 2019).

4. PERCEIVING FROM THE INSIDE OUT

82 **signals from each level:** Perceptual hierarchies are not strictly insulated from each other. A good rule of thumb is that there is more cross-modal interaction the further you get from the sensory periphery. See Felleman & Van Essen (1991); Stein & Meredith (1993).

83 **More recent experiments:** Grill-Spector & Malach (2004).

83 **computational theory of vision:** Marr (1982).

83 **impressive performance levels:** He et al. (2016).

84 **delightful exchange:** Anscombe (1959), p. 151, italics in original.

85 **"judgment and inference":** From *The Optics of Ibn al-Haytham*, c.1030, translated in Sabra (1989). The history is told in more detail in Jakob Hohwy's terrific *The Predictive Mind* (2013).

86 **Helmholtz saw himself:** Swanson (2016).

86 **those from well-to-do families didn't:** Bruner & Goodman (1947).

86 **Perceptual content, for Gregory:** Gregory (1980).

87 **theories differ in their details:** Clark (2013); Clark (2016); Hohwy (2013); Rao & Ballard (1999).

88 **adjusting top-down predictions:** Arguably, not all predictions are top-down, nor are prediction errors necessarily bottom-up. "Bottom-up predictions" can be thought of as constraints which reflect global and stable aspects of perceptual inference (Teufel & Fletcher, 2020). An example of a perceptual constraint could be the overrepresentation of vertical and horizontal orientations in natural images. The influence of shadows on luminance—discussed later—might also be a bottom-up predictive constraint.

91 **Dennett points out:** See Dennett (1998). Here's another example: In order to hear music, we don't need a little band in the head and an intricate system of intracranial microphones. Daniel Dennett is also responsible for the happy turn of phrase "mental figment."

91 **"color is the place":** Gasquet (1991). This quote has also been attributed to Paul Klee.

92 **agree about our hallucinations:** Many thanks to peer-reviewed rap artist Baba Brinkman for this provocative line about reality. There'll be more about the perception of "what is real" in chapter 6.

92 **short book on visual illusions:** *Eye Benders* (Gifford & Seth, 2013).

93 **these and a thousand other hypotheses:** Brainard & Hurlbert (2015); Witzel et al. (2017).

97 **changes what you consciously see:** A recent study by Biyu He and colleagues looked at how neural dynamics differ when a Mooney image is recognized, compared to when it is not. See Flounders et al. (2021).

97 **compelling auditory examples:** Chris Darwin has some excellent examples of sine wave speech online at www.lifesci.sussex.ac.uk/home/Chris_Darwin/SWS. I use another example in my 2017 TED talk: www.ted.com/talks/anil_seth_your_brain_hallucinates _your_conscious_reality. There are also auditory equivalents of The Dress. One example is a sound which some people hear as "Yanny" and others as "Laurel" (Pressnitzer et al., 2018). In 2020, a TikTok video appeared in which an ambiguous tinny noise from a cheap toy can be heard either as "green needle" or "brainstorm," depending on which words you are reading (time.com/5873627/green-needle-brainstorm-explained).

97 **perceptual experience is built:** See de Lange et al. (2018) for a review of experiments showing how expectations shape perception.

98 **Accurate—"veridical"—perception:** Another useful intuition pump for thinking about perceptual veridicality is peripheral vision—that part of the visual field away from the center of your gaze (foveal vision). The visual periphery has a much lower photoreceptor density than the fovea, yet visual experience in the periphery doesn't seem blurry. Does this mean that peripheral vision is less veridical than foveal vision, because it manifests an "illusion of sharpness" (or more specifically an illusion of "non-blurriness")? No! Sharpness and blurriness are properties of perceptual experience that are relative to the sensory data that each part of the visual system is tuned to. See Haun (2021) for an illuminating discussion, Lettvin (1976) for historical context, and Hoffman et al. (2015) for a wide-ranging discussion of perceptual (non-) veridicality.

98 **Locke proposed:** John Locke, *An Essay concerning Human Understanding* (1689), 14th ed. (1753). Although the example of color is intuitive, philosophers have energetically debated whether it really counts as a secondary quality (Byrne & Hilbert, 2011). A related distinction that appears in the philosophical literature is between different kinds of "kinds"—where the differences lie in the conditions necessary for something to exist. For example, money requires social conventions to exist and so is a "social kind." Water does not require social conventions to exist, and so is a "natural kind."

5. THE WIZARD OF ODDS

101 The Wizard of Odds: Thanks again to Baba Brinkman, who inspired this title. It comes from his *Rap Guide to Consciousness* (2018), for which I was a scientific adviser. See https://bababrinkman.com/shows/#consciousness.

102 see the lawn is wet: This example is adapted from F. V. Jensen (2000). For a thorough grounding in abductive reasoning, you can do no better than Peter Lipton's *Inference to the Best Explanation* (2004).

103 The rule itself: Here's how Bayes' rule is normally written: $p(H|D)=(p(D|H)*p(H))/(p(D))$. $p(H|D)$ is the posterior—the probability of the hypothesis H given the data D, $p(D|H)$ is the likelihood—the probability of the data given the hypothesis, and $p(H)$ and $p(D)$ are the prior probabilities of the hypothesis and the data, respectively. Note that coming up with a number for $p(D)$ can be difficult, but fortunately it is often not necessary. When the goal is to find the most likely posterior among a range of alternatives, as it often is, the $p(D)$s nicely cancel out.

104 wrongly conclude that you have a nasty disease: In 2009 the US government advised against mass mammographic breast cancer screening of women in their forties for exactly this reason. Mammograms at that time had about 80 percent sensitivity, which means they detected about 80 percent of women in this age group who had breast cancer when they were screened. But the tests also flagged cancer in about 10 percent of women who did not have the disease at the time. Critically, breast cancer has a low incidence—or "base rate"—of about 0.04 percent in this age group. Applying Bayes' rule, using this base rate as the prior, we can calculate that the probability of having breast cancer given a positive mammogram is only about 3 percent. Out of every 100 women testing positive, 97 will be cancer-free, and will undergo needless anxiety as well as potentially expensive and invasive additional investigation. One moral from this story is to improve the sensitivity and specificity of testing, and mammograms are indeed much better than they were. A recent UK study has suggested that screening women in their forties could now be worthwhile. See McGrayne (2012) and Duffy et al. (2020).

105 Bayesian inference has been applied: The surprisingly controversial history of Bayesian analysis is beautifully recounted in Sharon McGrayne's *The Theory That Would Not Die* (2012).

105 In the philosophy of science: For more on this, see Lakatos (1978) and Seth (2015b).

106 overturn my Bayesian belief: Hat tip to Paul Fletcher and Chris Frith, who made the same point about their Bayesian theory of hallucination and delusion in schizophrenia (Fletcher & Frith, 2009).

107 something has to happen: Strictly, X is a "random variable" because its value is determined by a probability distribution. The example in the figure on page 108 is a "continuous" probability distribution (also called a probability density function), since X can take any value within the allowed range. If X could take only specific values—for example "heads" or "tails"—we would have a "discrete" probability distribution.

109 the mean signifies the probability: There are two ways to interpret mappings between Bayesian beliefs and the brain. The weaker version is that we, as external observers, use these beliefs to represent things to us, just as we might use a physical map to represent our surroundings. In this view, observed neural activity might represent certain states of affairs to us, as scientists. The stronger interpretation is that the brain uses these beliefs (or something like them) to represent things to itself. This second interpretation exemplifies the "Bayesian brain" hypothesis and is central to the idea of the brain as a prediction machine. Failure to distinguish these two meanings of "representation" has been the source of much confusion throughout cognitive science and neuroscience. See Harvey (2008).

111 all the top-down predictions: Some people refer to the bottom-up and top-down pathways as, respectively, "feedforward" and "feedback." From the perspective of predictive processing, this is the wrong way around. In engineering, "feedback" is usually associated with an error signal which is used to adjust the "feedforward" control signal. Therefore, in predictive processing, the bottom-up connections are now the "feedback" connections, because these connections convey the error signals. Complicating matters further, as mentioned earlier, some (global, stable) predictions may be embedded in bottom-up signals, with top-down connections now conveying prediction errors (Teufel & Fletcher, 2020).

111 most prominently *predictive processing*: I use predictive processing in this book as a shorthand for a variety of theories, including but not limited to predictive coding, which include the core mechanism of prediction error minimization. In doing so, I do not mean to diminish their differences, which are interesting and important (Hohwy, 2020a).

112 tries to minimize these errors: In Bayesian terms, a generative model is specified by a combination of a prior and a likelihood. Think of it as "the joint probability of the hypothesis and the data together," $p(H,D)$. Putting things this way mathematically underwrites the claim that prediction error minimization approximates Bayesian inference (Buckley et al., 2017).

112 cascade further down: Hierarchical prediction error minimization suggests that the parts of the brain involved in perception should be richly endowed with top-down connectivity conveying signals from higher hierarchical levels to lower levels. Many studies have shown this to be the case; see, for example, Markov et al. (2014). The presence of such rich top-down connectivity is harder to explain from a bottom-up perspective on perception.

113 precision weighting: See Feldman & Friston (2010). Precision weighting is achieved by altering priors on precisions—so-called hyperpriors—so that the inferred precision is increased or decreased.

114 video demonstration: See also Simons & Chabris (1999).

115 as if from thin air: See Kuhn, Amlani, & Rensink (2008) for a review of psychology and magic.

116 to act effectively: My University of Sussex colleague Andy Clark has long championed "action-oriented" formulations of predictive processing. His book *Surfing Uncertainty* (2016) is a landmark resource.

116 generating actions: As the neuroscientist György Buzsáki argues in his book *The Brain from Inside Out*, this perspective poses challenges, and opens new opportunities, for experimental neuroscience. Most experimentalists, though by no means all, study the brain by examining its activity in response to external stimulation, rather than as an intrinsically dynamic, active system. See Buzsáki (2019), and also Brembs (2020).

117 called "active inference": Friston et al. (2010)

118 virtuous circle: Alexander Tschantz, Christopher Buckley, Beren Millidge, and I have been developing new machine learning algorithms on this basis which are able to learn generative models from small amounts of data (Tschantz et al., 2020a). Interestingly, the possibility of "bottom-up" predictions (Teufel & Fletcher, 2020) has a promising application in this context. It can be related to the powerful technique of "amortization" in machine learning, where an approximate Bayesian posterior is computed through a feed-forward (bottom-up) sweep through an appropriately trained artificial neural network.

119 **"disattention"**: This idea can be experimentally tested, and it indeed turns out that during action, proprioceptive sensory sensitivity is reduced, just as would be expected (C. E. Palmer et al., 2016). The sensory attenuation accompanying action also provides a neat explanation for why we can't tickle ourselves (Brown et al., 2013).

6. THE BEHOLDER'S SHARE

121 **"the age of insight"**: Kandel (2012).

122 **later popularized**: Gombrich (1961).

122 **contributed by the perceiver**: Seth (2019b).

122 **As Kandel put it**: Kandel (2012), p. 204.

122 *Hoarfrost at Ennery*: This example is taken from Albright (2012) as adapted in Seth (2019b). See www.wikiart.org/en/camille-pissarro/hoarfrost-1873.

122 **"palette-scrapings"**: The phrase appears in Leroy's satirical review of Impressionist art published in *Le Charivari* on 25 April 1874—a review which coined the term "Impressionism."

122 **"innocent eye"**: Gombrich (1961).

123 **paintings become experiments**: In his later years, Pissarro suffered from serious eye problems. His fellow Impressionists Claude Monet and Edgar Degas suffered similarly. It is intriguing to consider whether, and how, their impaired eyesight contributed to their artistic insight. While some influence is plausible—perhaps they were more sensitive to patterns of light rather than details of objects—a cautionary tale comes from another painter, El Greco (1541–1614). El Greco's works often included unnaturally elongated figures, a feature which was attributed by some to his pronounced astigmatism. In this story, he painted elongated figures because that's what he saw. But psychologists correctly pointed out that by this logic, he would also have seen his canvas as elongated, canceling out any effect of astigmatism. This sort of logical error has been called the "El Greco fallacy," and it still catches out perception researchers even today. See Firestone (2013).

123 **"When we say"**: Gombrich (1961), p. 170.

123 **the nature of experience**: Gombrich's insight was later echoed by the writer, critic, and artist John Berger, whose influential 1972 book *Ways of Seeing* opened with the line "The relation between what we see and what we know is never settled." Compared to the culturally conservative Gombrich, Berger emphasized the political and cultural influences on perception, highlighting how what we see may differ among people and between groups (Berger, 1972).

123 **One experimental prediction**: In fact, this prediction is not all that straightforward; see Press et al. (2020).

124 **"continuous flash suppression"**: This is a variant of the better-known method of "binocular rivalry" which I described in chapter 1 (Blake et al., 2014).

124 **Our experiment**: Pinto et al. (2015). Our study was more complicated than summarized here. We ran a large number of control studies to rule out, as best we could, other factors

like biases in how people make responses, or in how they focus their attention, in accounting for our results. See de Lange et al. (2018) for a review of other, similar studies, and Melloni et al. (2011) for an influential early contribution to this literature.

126 **a powerful technique:** "Brain reading" involves training machine learning algorithms to classify brain activity into different categories. See Heilbron et al. (2020).

127 **see faces in things:** www.boredpanda.com/objects-with-faces.

129 **build a "hallucination machine":** Suzuki et al. (2017).

129 **Networks like this:** Specifically, the networks are deep convolutional neural networks (DCNNs) which can be trained using standard backpropagation algorithms. See Richards et al. (2019).

130 **reverses the procedure:** In the standard "forward" mode, an image is presented to the network, activity is propagated upward through the layers, and the network's output tells us what it "thinks" is in the image. In the deep dream algorithm—and in Keisuke's adaptation—this process is reversed. We fix the network *output* and adjust the *input* until the network settles into a stable state. See Suzuki et al. (2017) for details.

130 **much more compelling:** The full hallucination machine experience—in which a panoramic video is viewed through a head-mounted display—is considerably more immersive than any still image. Sample movie footage is here: www.youtube.com/watch?v=TlMBnCrZZYY.

131 **are the way they are:** When these computational models map onto hypotheses about neural circuits, this approach can be called "computational *neuro*phenomenology"—a computationally boosted version of the neurophenomenology that traces back to Francisco Varela (1996). In this spirit, Keisuke Suzuki, David Schwartzman, and I are developing new versions of the hallucination machine which explicitly incorporate generative models, and which therefore map more closely onto what we believe is happening in real brains. Our new hallucination machines are able to capture a much wider variety of hallucinatory experience than the original.

133 **exploring the principles of objecthood:** Seth (2019b).

133 **the painter investigates:** Merleau-Ponty (1964).

133 **by "sensorimotor contingency theory":** See O'Regan (2011); O'Regan & Noë (2001). Like all theories, this builds on a great deal of prior work, notably in this case James Gibson's and Maurice Merleau-Ponty's ideas about how perception depends on embodied action, as well as the philosophical phenomenology of Edmund Husserl and Maurice Merleau-Ponty. Gibson's notion of "affordances"—which we'll meet again in chapter 9—captures the idea that we perceive objects in terms of the possibilities for behavior that they afford (Gibson, 1979). Husserl proposed that "perception has horizons made up of other possibilities of perception, as perceptions that we could have, if we actively directed the course of perception otherwise" (Husserl, 1960 [1931], sec. 19)). Merleau-Ponty, who was strongly influenced by Husserl, emphasized the embodied aspects of perceptual experience, and his *The Phenomenology of Perception* remains influential (Merleau-Ponty, 1962).

134 **in 2014, I proposed:** Seth (2014b). This is another example of "computational phenomenology." See also Cisek (2007) for a related neurophysiological theory in terms of "affordance competition."

135 **"grapheme-color synesthesia":** Seth (2014b).

135 **one recent experiment:** Suzuki et al. (2019).

137 **snake-filled image:** Another good example is provided by motion aftereffects, such as the waterfall illusion. Stare straight at a waterfall (or at a video of a waterfall) for a while, then

look away, perhaps at a rock face next to it. The rock face will appear to move upward, while also appearing to remain in the same place.

138 powerful video example: See www.youtube.com/watch?v=hhXZng6o6Dk.

139 experiences of time are controlled hallucinations: Many time-focused neuroscientists will disagree with this view. In fact, most psychological and neural models of time perception assume some kind of neuronally implemented "pacemaker" which serves as a benchmark against which physical time can be compared, giving rise to perceptions of duration (van Rijn et al., 2014). Others think that time perception depends on clock-like signals coming from the body (heart rate and so on)—see Wittmann (2013)—but our research casts doubt on this idea too (Suárez-Pinilla et al., 2019).

139 characteristic biases: The underestimation of long durations and overestimation of short durations is an example of a "regression to the mean" effect. This effect is seen in many if not all perceptual modalities, and is a signature of Bayesian inference since the mean can be thought of as a prior. In time perception, this effect is known as Vierordt's law.

140 showing the same biases: See Roseboom et al. (2019). Further support for Warrick's idea came from the finding that the match between the computational model and human performance was even closer when the network input was limited to the parts of each video that a person was looking at.

140 we used fMRI: Sherman et al. (2020).

140 My favorite, by a distance: Stetson et al. (2007). There was of course a large net at the bottom, to catch the volunteers.

141 "substitutional reality": For an early version of this project, see Suzuki et al. (2012).

142 we can now test ideas: Phillips et al. (2001). There are potentially important sociological implications for this line of research. To the extent that people experience their perceptions as being both "real" and veridical, it will be difficult to accept that others might have different perceptual experiences, even when faced with the same objective circumstances. This is one reason why there was such a hullabaloo about The Dress (see chapter 4). People who saw it one way simply could not accept that other ways of seeing it were possible, precisely because they experienced their perceptions as directly revealing an objective reality. This kind of perceptual drift—a generalization of the echo chambers of social media—has many implications for how we might recognize and resolve or accommodate differences among individuals, groups, and cultures (Seth, 2019c).

143 As Hume put it: These quotes trace back to Hume's *Treatise of Human Nature* (1738, 1:3.14:25) and *Inquiry Concerning the Principles of Morals* (1751, Appendix 1:19), as found in Kail (2007), p. 20. I encountered the material in Dennett (2015).

144 we perceive *with* and *through*: Philosophers call this "transparency" (Metzinger, 2003b).

7. DELIRIUM

151 Up to a third: Collier (2012).

151 severe long-term consequences: Davis et al. (2017).

8. EXPECT YOURSELF

154 "teletransportation paradox": This thought experiment has been independently attributed to the philosopher Derek Parfit and the author Stanislaw Lem.

156 "bundle" of perceptions: This view of the self has been immortalized in philosophy as "bundle theory."

156 very brilliant book: Metzinger (2003a).

156 one craniopagus twin can feel: This is the case of Krista and Tatiana Hogan, from British Columbia in Canada: www.nytimes.com/2011/05/29/magazine/could-conjoined-twins -share-a-mind.html.

160 can come undone: See Brugger & Lenggenhager (2014) for a review of disorders of the bodily self.

160 behind the windows of the eyes: For a classic first-person deconstruction of this intuition, see Douglas Harding's *On Having No Head* (1961).

161 cornerstone of research: The original paper describing the rubber hand illusion was published in 1998 (Botvinick & Cohen, 1998) and has spawned a small industry of follow-up studies. People have looked at whether three-hand or no-hand illusions are possible, at whether inducing ownership of hands with different skin colors can alter implicit racial biases, and even at whether mice are susceptible to a "rubber tail" illusion. See Braun et al. (2018) and Riemer et al. (2019) for reviews.

163 extended to the entire body: The experiment also worked when volunteers saw the back of a fake, computer-generated body—an "avatar"—rather than their own body. See Ehrsson (2007); Lenggenhager et al. (2007).

164 inscribed in history and culture: Monroe (1971).

164 "I have a queer sensation": Quoted in Tong (2003), p. 105.

164 "I see myself lying": Blanke et al. (2004), p. 248.

164 The common factor: Blanke et al. (2015).

165 evidence for the malleability: See Brugger & Lenggenhager (2014).

165 "body swap" illusion: Petkova & Ehrsson (2008).

166 The aim of BeAnotherLab: www.themachinetobeanother.org.

168 highlighted by a recent study: Lush et al. (2020). In this paper, we use the term "phenomenological control" rather than "hypnosis," partly because "hypnosis" carries unfortunate historical baggage. The idea that implicit expectations arising from experimental design can affect participants' experiences and behavior is a well-known, though often insufficiently appreciated, issue in psychology, tracing back to early work on so-called "demand characteristics" (Orne, 1962). Importantly, simply contrasting synchronous versus asynchronous stroking is not enough to control for demand characteristics in the rubber hand illusion, because people have strong expectations about what to experience in these different conditions (Lush, 2020). For a summary of our studies on the rubber hand illusion, phenomenological control, and demand characteristics, see Seth et al. (2021).

168 hypnotic suggestion: As well as inducing behaviors and reports of subjective experiences, hypnotic suggestion can generate physiological and neurophysiological responses (or suppress them, in the case of hypnotic analgesia) (Barber, 1961; M. P. Jensen et al., 2017; Stoelb et al., 2009). These ostensibly more objective measures may therefore also be confounded by suggestibility. Fortunately, all this can be seen as an opportunity, not just as a problem. Hypnotic suggestion provides a powerful method for studying how top-down expectations can generate, or extinguish, perceptual experience. We are currently running a number of experiments to look more deeply into which kinds of perceptual experience—of the self and of the world—can be shaped or generated by suggestion effects.

169 truly *self-aware*: The Estonian-Canadian psychologist Endel Tulving has called this kind of self-awareness "autonoetic consciousness" (Tulving, 1985).

170 **His diaries:** Diary entries from Deborah Wearing's *Forever Today* (2005).

171 **"Clive was under the constant impression":** Wearing (2005), pp. 202–3.

172 **"He no longer has":** www.newyorker.com/magazine/2007/09/24/the-abyss.

173 **Much of our social perception:** The existence of direct social perception, defined this way, is not universally accepted. Other approaches propose instead that our awareness of others' mental states is inferred from their behavior in a manner that is distinct from perception. See Gallagher (2008) and C. J. Palmer et al. (2015) for more on this.

174 **Active inference in social perception:** An implication of social active inference is that, just like visual active inference, the underlying generative models encode conditional predictions about the consequences of actions. In vision, as we saw in chapter 6, these predictions are about how visual sensory signals would change given this-or-that action. There, I argued that these conditional predictions underlie the phenomenological property of "objecthood." In a 2015 paper, Colin Palmer, Jakob Hohwy, and I proposed that something similar happens in social perception. Our idea is that others' mental states seem "real" to the extent that our brain encodes a rich repertoire of conditional predictions about how they may change given this-or-that action. For example, such a prediction may be about how someone's belief or emotional state would change given a particular utterance (like "go get me some wine"). This idea provides a prediction machine gloss on what it means to have a theory of mind, and it may provide a useful way of interpreting apparent deficits in social perception, as happens for example in autism. See C. J. Palmer et al. (2015).

174 **socially nested predictive perceptions:** Neuroscientific discussions of social perception often highlight so-called "mirror neurons." These neurons—first discovered in monkey brains by the Italian neuroscientist Giacomo Rizzolatti and his colleagues—fire away both when an animal acts and when it observes the same action performed by another animal (Gallese et al., 1996). The neurons are said to "mirror" the behavior of other animals, because they respond as though the observer itself were acting. The ability of these neurons to respond in this way has been proposed as a foundation for all manner of social phenomena. Such proposals, however, place a heavy explanatory burden on what in the end are just specific types of brain cell. They commit the same error of simplification that people make when, on the basis of fMRI scans, they may say that activity in this-or-that region explains "love" or "language." See Caramazza et al. (2014).

175 **"no man is an island":** "No man is an island, entire of itself; every man is a piece of the continent, a part of the main" (Donne, 1839), pp. 574–75. The psychologist Chris Frith has taken this view even further, arguing that the primary function of all conscious experience is social (Frith, 2007).

175 **persisting from moment to moment:** Antonio Damasio highlights this aspect of selfhood in his book *The Feeling of What Happens* (Damasio, 2000).

176 **"Contrary to the perception":** James (1890), p. 242.

9. BEING A BEAST MACHINE

178 **"We do not see things":** Anaïs Nin, *The Seduction of the Minotaur* (1961), p. 124. Nin attributed the remark to an ancient Talmudic text.

178 **In the Great Chain:** The concept of the Great Chain of Being originated in Ancient Greece with Plato, Aristotle, and Plotinus, and became fully developed in western Europe during the middle ages.

179 annoy the powerful Catholic Church: For a vivid historical account, see George Makari's *Soul Machine* (2016).

179 prove the existence of a benevolent God: https://en.m.wikipedia.org/wiki/Trademark _argument. See also Hatfield (2002).

180 "The bodies of both man and beast": Shugg (1968), p. 279. For the original, see *The Philosophical Works of Descartes*, trans. E. S. Haldane and G. R. T. Ross (New York, 1955), vol. 1, 114–16, 118.

181 *l'homme machine* (man machine): La Mettrie (1748).

181 whether life and mind are continuous: Godfrey-Smith (1996); Maturana & Varela (1980).

182 "sense of the internal": Craig (2002).

182 Interoceptive sensory signals: Critchley & Harrison (2013).

183 insular cortex: See Barrett & Simmons (2015) and Craig (2009) for more on the role of the insular cortex in interoception.

183 "We feel sorry": James (1884), p. 190. The debate between "classical" emotion theorists, who follow Darwin in proposing innate emotions conserved across species, and "constructivists," who do not, continues to this day. In the former camp we have the biologist Jaak Panksepp and his followers. Panksepp argued that a set of basic emotions are instantiated by specific (and evolutionarily old) neural circuits (Panksepp, 2004); see also Darwin (1872). The latter camp is exemplified by the neuroscientists Lisa Feldman Barrett and Joe LeDoux, who propose different versions of the idea that human emotions depend on cognitive evaluations—a perspective, as we will see, similar to my own. For more on the history of emotion theories, see Barrett & Satpute (2019) and LeDoux (2012).

184 still debated: See for example Harrison et al. (2010).

184 "appraisal theories": Schachter & Singer (1962) is the classic reference for appraisal theories.

184 an inventive study: Dutton & Aron (1974).

186 it occurred to me: The outlines of this idea first appeared in a 2011 paper (Seth et al., 2011b), and were refined in a 2013 paper which has since become the standard reference (Seth, 2013). The core ideas were extended from 2015 onward into the "beast machine" theory of consciousness and self: Seth (2015a); Seth (2019a); Seth & Friston (2016); Seth & Tsakiris (2018).

187 predictions about the causes of interoceptive signals: Damasio's ideas deeply shaped my own thinking, especially when it comes to the self; see Damasio (1994, 2000, 2010). Lisa Feldman Barrett, like me, emphasizes the role of interoceptive predictions in emotion— see Barrett (2017b) and Barrett & Simmons (2015), as well as her excellent book *How Emotions Are Made* (Barrett, 2017a).

187 "heartbeat evoked potentials": See Petzschner et al. (2019). Additional evidence for interoceptive inference is beginning to accrue from animal studies. For example, two recent experiments have suggested that neurons in the insular cortex of mice encode something like interoceptive predictions (Gehrlach et al., 2019; Livneh et al., 2020).

188 flashed in time with their heartbeat: See Aspell et al. (2013) and Suzuki et al. (2013). In a related study, Micah Allen and colleagues showed that unexpected physiological arousal influences perception of visual stimuli—again suggesting an interaction between exteroceptive and interoceptive processes (Allen et al., 2016). See Park & Blanke (2019) for a review, and Brener & Ring (2016) and Zamariola, Maurage, Luminet, & Corneille

(2018) for discussions of problems associated with measuring interoceptive sensitivity through heartbeat detection. The influence of hypnotic suggestibility on embodiment was discussed in chapter 8.

189 "the scientific study of control": Wiener (1948).

190 valuable insights: The history of cybernetics shows not only how academic disciplines which we now consider as being distinct were once part of a common approach, but also reveals how science can sometimes meander away from paths that might have delivered extraordinary riches had they been followed. For more on this, I recommend Jean-Pierre Dupuy's *On the Origins of Cognitive Science: The Mechanization of Mind* (2009).

190 "Good Regulator Theorem": Conant & Ashby (1970).

190 "must be a model": See Seth (2015a) and Seth & Tsakiris (2018) for more on the distinction between being a model and having a model.

192 about finding out things or about controlling things: In research papers, I refer to the distinction as being between *epistemic* (exploratory, information-seeking) and *instrumental* (goal-directed, control-oriented) forms of predictive perception: Seth (2019a); Seth & Tsakiris (2018); Tschantz et al. (2020b).

192 of an *essential variable*: Ashby (1952). For example, human core body temperature must remain between 32°C (89.6°F) and 40°C (104°F), otherwise death rapidly follows.

192 these actions can be both: External and internal actions are distinguished by the type of muscle involved. External actions depend on the skeletal (striated) muscle system, while internal (intero-)actions depend on the visceral (smooth) and cardiac muscle systems. These muscle types are in turn controlled by different branches of the peripheral nervous system (the part of the nervous system that lies outside the brain and spinal cord). Skeletal muscles are controlled by the somatic branch, and visceral and cardiac muscles are controlled by the autonomic branch.

194 "optic acceleration cancellation": See McLeod et al. (2003).

194 "affordances": Gibson (1979).

194 "perceptual control theory": This theory is often summarized by the slogan that control systems control what they *sense*, not what they *do* (Powers, 1973). See Marken & Mansell (2013) for a more recent articulation of the theory.

195 There is no phenomenology: I'm not against the metaphorical ascription of shapes to nonvisual experiences. Pains can be sharp or dull. Some tastes are also sharp, and some emotions can be described this way too—a pang of jealousy, perhaps. But these experiences do not have shapes in the same way that experienced cups, cats, and coffee tables have shapes.

197 "allostasis": Sterling (2012). See Tschantz et al. (2021) and Stephan et al (2016) for computational models of allostatic interoceptive control.

198 the beast machine theory: See Seth (2013), Seth (2014b), Seth (2015a), Seth (2019a), Seth & Friston (2016), and Seth & Tsakiris (2018) for more technical versions of the beast machine theory and its components. The theory has many ancestors and influences, which I cannot do full justice to here. These include Thomas Metzinger's philosophical examination of the self (Metzinger, 2003a), and the landmark accounts of predictive processing from Andy Clark and Jakob Hohwy (Clark, 2016; Hohwy, 2013). The theory owes a particular debt to other proposals of deep—but different—links among life, body, mind, and consciousness. Here, I have been strongly influenced by Antonio Damasio (e.g., 1994, 2010), Gerald Edelman (e.g., 1989), Karl Friston (e.g., 2010), Joe LeDoux (e.g., 2019), and Evan Thompson (e.g., 2014; see also Varela et al., 1993). For related ideas, see Panksepp (2005), Park & Tallon-Baudry (2014), Solms (2021), Metzinger's concept of "existence bias" (Metzinger, 2021), and the work of Lisa Feldman Barrett (e.g., 2017b).

198 **conscious beast machines:** A close association between consciousness and physiological regulation raises new questions about the role of the brainstem—the set of nuclei lying between the deepest parts of the cerebral hemispheres and the spinal cord. Typically, the brainstem has been thought of as an "enabling factor" for consciousness, much like a power cable is an enabling factor for a TV. But the brainstem plays a highly active role in physiological regulation, leading some to suggest that this is where consciousness arises—with no need for cortex (Solms, 2021; see also Merker, 2007). I think this is extremely unlikely, given the weight of explanatory evidence linking cortex (and thalamus) to conscious states. Having said this, the brainstem may well play a more decisive role in shaping conscious states than suggested by the power cable analogy—see Parvizi & Damasio (2001) for a nuanced view.

200 **systematically *mis*perceiving:** It is interesting to consider whether self-change-blindness is attenuated during illness or injury, when it might be useful for the brain to more accurately perceive what's going on in the body. There is a new subfield of cognitive neuroscience which deals with questions like this, called "computational psychosomatics" (Petzschner et al., 2017).

200 **not in order to know ourselves:** The Greeks had this down already. While Socrates is associated with the phrase "know thyself," the Stoics emphasized the importance of equanimity and self-control. Advocates of perceptual control theory might go further still, to say that we regulate our physiological condition in order to perceive ourselves as stable.

201 **a rare delusion:** Cotard (1880).

201 **gone significantly awry:** One possible way the self may lose its reality is by the underlying generative models failing to encode a rich repertoire of conditional or counterfactual interoceptive predictions concerning how actions affect physiological regulation. This is by analogy with how visual conditional predictions may underlie the phenomenology of "objecthood." See Seth & Tsakiris (2018).

201 **something special about flesh:** The claim that consciousness depends on a specifically *biological* property (that a computer made of silicon could never have) is sometimes called *biological naturalism*. I do not use this term here because it has been employed in different ways by different people—see the discussion in Schneider (2019).

10. A FISH IN WATER

204 **as has yet been proposed:** Friston has published a large number of papers on the free energy principle. Two key overviews are Friston (2009, 2010).

204 **our own review:** Buckley et al. (2017).

204 **struggles to comprehend:** www.lesswrong.com/posts/wpZJvgQ4HvJE2bysy/god-help-us-let-s-try-to-understand-friston-on-free-energy. Other gems include "How to read Karl Friston (in the original Greek)" by Alianna Maren, www.aliannajmaren.com/2017/07/27/how-to-read-karl-friston-in-the-original-greek, and "Free Energy: How the f*ck does that work, ecologically" by Andrew Wilson and Sabrina Golonka, psychsciencenotes.blogspot.com/2016/11/free-energy-how-fck-does-that-work.html.

205 maintain their boundaries: The FEP talks about boundaries in terms of "Markov blankets"—a concept from statistics and machine learning. For a system described by a set of random variables, a Markov blanket is a statistical partitioning of the system into "internal states," "external states," and "blanket states," with the blanket separating the internal from the external. Markov blankets satisfy the requirement that variables inside the blanket (internal states) are conditionally independent of those outside the blanket (external states), and vice versa. This means that the dynamics of internal states can be fully predicted from past internal states and blanket states. See Kirchhoff et al. (2018) for more on Markov blankets and the FEP, and Bruineberg et al. (2020) for an illuminating criticism.

208 approximates sensory entropy: Technically, free energy provides an upper bound on a quantity called "surprisal" or "self-information," which can be thought of as specifying how (statistically) unexpected an event is. The upper bound means that free energy cannot be less than surprisal. Surprisal is related to the information-theoretic quantity of entropy in that, under nonequilibrium steady-state assumptions, the long-term average of surprisal is entropy. Informally, entropy is like uncertainty—and uncertainty is the average surprise you expect to encounter.

209 after some mathematical juggling: More details for the mathematically inclined. Free energy is defined in terms of two probability distributions: (i) a *recognition density* which encodes a current best guess about the state of the environment, and (ii) a *generative density* which encodes a probabilistic model of how environmental states shape (generate) sensory inputs. Here, "environment" refers to the hidden causes of sensory signals, whatever they may be. Free energy quantities have two components: an *energy* that corresponds to surprisal, and a *relative entropy* that reflects how "far apart" the recognition density is from the true posterior density (the probability over states of the environment, given sensory inputs). The distance between these densities is measured by a quantity which in information theory is called the Kullback–Leibler (KL) divergence. If the recognition and generative densities are assumed to be Gaussian (and given some other assumptions, for example about independence of timescales), then free energy maps directly onto precision-weighted prediction error in predictive processing. Because the smallest measure of "far apart" is zero, this means that free energy is always bigger than surprisal (i.e., it provides an upper bound). This in turn means that decreasing free energy must either reduce the divergence between the recognition density and the true posterior (coming up with better perceptual inferences) or reduce surprisal (by sampling new sensory inputs).

209 Paraphrasing Friston: Friston (2010). Jakob Hohwy has neatly termed this process "self-evidencing" (Hohwy, 2014). Mathematically, this view is justified because minimizing free energy is equivalent to maximizing (Bayesian) model evidence. Indeed, just as free energy provides an upper bound on surprisal, it provides a lower bound on model evidence (the so-called evidence lower bound, or ELBO, in machine learning); see Winn & Bishop (2005).

210 out of the dark room: The "dark room problem" was one of the objections first raised against the FEP (Friston et al., 2012). It resurfaced again while I was writing this book, and my colleagues and I rebutted it again. See Seth et al. (2020); Sun & Firestone (2020).

211 FEP will be judged: Hohwy (2020b), p. 9.

212 textbook on statistical mechanics: Goodstein (1985), p. 1.

212 **more integrative, and more powerful:** One example of this way of augmenting the beast machine theory takes us back to the notion that living systems maintain a boundary between themselves and their environment—where a boundary in the FEP is understood in terms of a Markov blanket (see note to page 205). For Friston, the mere existence or identification of a Markov blanket directly implies that active inference is happening. See Kirchhoff et al. (2018) and again Bruineberg et al. (2020) for a critique.

213 **enable us to do better experiments:** One example is our research exploring how free energy-minimizing agents learn adaptively biased perceptual models of their environments (Tschantz et al., 2020b).

213 **theories *for* consciousness science:** Hohwy & Seth (2020). There have been some other attempts to link the FEP to consciousness, for example in terms of the temporal depth of generative models (Friston, 2018); see also Solms (2018); Solms (2021); Williford et al. (2018).

215 **attempts underway:** These attempts are in the form of an "adversarial collaboration" in which proponents of the two theories sign off in advance about whether the outcome of an experiment will support or undermine their preferred theory. This particular adversarial collaboration pits IIT against active inference (not the FEP itself). I mentioned one of the proposed experiments in chapter 3: IIT predicts that inactivating already inactive neurons will make a difference to conscious perception, whereas active inference does not.

11. DEGREES OF FREEDOM

217 **"She bent her finger":** McEwan (2000). Hat tip to Patrick Haggard of University College London for this quote.

218 **There is not even clarity:** For an adroit guide to this philosophical minefield, see Bayne (2008).

218 **"radical, absolute, buck-stopping":** Strawson (2008), p. 367.

219 **can chug along:** This view, which I subscribe to, expresses what philosophers call "compatibilism." Compatibilism holds that some reasonable conceptions of free will are compatible with the universe being deterministic. In contrast, "libertarian" free will (nothing to do with the political philosophy) is the philosophical version of spooky free will. There are also proponents of so-called "hard determinism," which holds that determinism is true and concludes from this that no reasonable conception of free will can survive. Note that I can be agnostic about whether the universe is deterministic, which I am, and still be a compatibilist.

220 **neither *feel* random, nor *are*:** Even if the universe is deterministic at some fundamental level, it could be that apparently random fluctuations at the level of neurons and synapses play significant roles in brain function. This is possible, perhaps even likely, but again it doesn't matter.

220 **well-known phenomenon:** The readiness potential was first recorded in the 1960s by the German physiologists Hans Kornhuber and Lüder Deecke, who called it the *Bereitschaftspotential* (Kornhuber & Deecke, 1965).

220 **experienced the "urge":** For philosophers, intentions and urges are different things. I might feel the *urge* to hit someone who is annoying me, but I will inhibit this urge because I do not *intend* to hurt anyone, or get myself arrested. Intentions are subject to reasons and norms in a way that urges are not. In simple cases like voluntarily flexing a finger, they amount to much the same thing, and Libet himself mostly used the terms interchangeably.

221 has already started: Libet (1985). Unsurprisingly, both the readiness potential onset and the timing of the conscious intention happen prior to the movement itself.

222 "free won't": Libet et al. (1983).

222 we don't see it: The philosopher Al Mele had made a similar conceptual point a few years earlier (Mele, 2009).

223 little sign of anything: Schurger et al. (2012).

224 if I am being forced: See Caspar et al. (2016) for an interesting experiment on this issue.

225 "Man can do what he wills": This quote originates in an essay that Schopenhauer presented to the Royal Norwegian Society of Sciences in 1839. For a translation, see Zucker (2013), p. 531. In Schopenhauer's distinction, one can see the roots of all addiction.

226 Here's where "degrees of freedom": There is an interesting connection to cybernetics here. Ross Ashby, who we met in chapter 9 for his Good Regulator Theorem and concept of an essential variable, is also known for his earlier "Law of Requisite Variety" (Ashby, 1956). This law, or principle, states that a successful control system must be capable of entering at least as many states as the environment which is perturbing it. As Ashby put it, "Only variety can force down variety." See Seth (2015a).

226 Voluntary behavior depends: Dennett (1984).

227 three processes: Brass & Haggard (2008); Haggard (2008); Haggard (2019).

227 stimulation of these regions: Fried et al. (1991). More intense stimulation to this area can generate both an urge and the corresponding action.

227 localizable to more frontal parts: Brass & Haggard (2007).

229 *The Illusion of Conscious Will*: Wegner (2002).

230 a problem that doesn't exist: As Sam Harris put it on a recent podcast, "The problem is not merely that the problem of free will makes no sense objectively, it makes no sense subjectively, either." See https://samharris.org/podcasts/241-final-thoughts-on -free-will/.

230 being "in the moment": Csikszentmihalyi (1990); Harris (2012).

231 Brain injuries, or unlucky draws: Della Sala et al. (1991); Formisano et al. (2011).

231 awkwardly located brain tumor: The case of tumor-induced pedophilia is described in Burns & Swerdlow (2003). That of Charles Whitman has been related many times, including in a 2011 article in *The Atlantic* by David Eagleman: www.theatlantic.com/mag azine/archive/2011/07/the-brain-on-trial/308520.

231 such as the philosopher Bruce Waller: See Waller (2011) for the view that moral responsibility is incoherent, and Dennett (1984, 2003) for the alternative. There is an illuminating recent debate on this issue in Dennett & Caruso (2021).

231 Einstein stated: "I claim credit for nothing. Everything is determined, the beginning as well as the end, by forces over which we have no control. It is determined for the insect as well as for the star. Human beings, vegetables or cosmic dust, we all dance to a mysterious tune." In this passage, Einstein's well-known commitment to determinism is on full display. (He also famously said that "God does not play dice with the universe"—a rejection of the inherent randomness of quantum mechanics.) However, as we've seen, one doesn't need to embrace determinism in order to reject spooky free will. The passage comes from an interview conducted by George Sylvester Viereck, published in the *Saturday Evening Post* on 26 October 1929 (p. 117).

232 among the animals: The neurobiologist Björn Brembs has argued that traces of what we call free will can be found even in very simple creatures, when they behave in variable and apparently spontaneous ways that also appear to come "from within." Because these behaviors—such as the unpredictable "escape response" of a cockroach—are beneficial for avoiding predators, perhaps they reflect the evolutionary origins of how we humans have come to control our many degrees of freedom. See Brembs (2011).

12. BEYOND HUMAN

235 history of animal criminal prosecution: Evans (1906).

237 animal-robots of Cartesian dualism: Descartes's views on animals were published in the seventeenth century and so partly coexisted with these medieval beliefs and practices. The Cartesian view became dominant later on, as the Enlightenment took hold across Europe.

237 absence of so-called "high-level" cognitive abilities: Advocates of "higher-order thought" theories of consciousness (which I described briefly in chapter 1) might disagree with this; see Brown, Lau, & LeDoux (2019).

237 myopia exemplified by the Cartesian view: At the opposite extreme to Descartes, Darwin adopted a strongly anthropomorphic perspective, especially regarding the expression of emotions in animals (Darwin, 1872). His assumption of a conserved set of emotions across many species licensed the use of animal experiments to study human emotions, a line of work later taken up by Jaak Panksepp and others, and became baked into contemporary culture in the form of "hard-wired" or "basic" emotions associated with specific facial expressions, as in the work of Paul Ekman (Ekman, 1992). As explained in chapter 9, this "basic emotion" view has been challenged by contemporary "constructivists" such as Lisa Feldman Barrett and Joe LeDoux, who, like me, emphasize the role of top-down interpretations in forming conscious content. See Barrett (2017b); LeDoux (2012).

238 Even rats, one study suggests: See Steiner & Redish (2014), p. 1001.

239 wherever there is life: The biologist Ernst Haeckel coined the term "biopsychism" in 1892 to describe the view that all and only living things are sentient—as distinct from the panpsychist view that consciousness is a property of all forms of matter. See Thompson (2007).

239 set aside raw brain size: A more sophisticated measure than raw brain size is the "encephalization quotient," which is a measure of relative brain size, taking body size into account. There is much discussion about whether the encephalization quotient is a reliable predictor of cognitive ability across species (Herculano-Houzel, 2016; Reep et al., 2007). However, there is little reason to take either brain size or encephalization quotient as a marker of the presence of consciousness across species.

239 seventeen distinct properties: Seth et al. (2005).

240 similar effects across mammalian species: Kelz & Mashour (2019).

240 though seals do so only: Lyamin et al. (2018); Walker (2017).

241 distinctive inner universe: Uexküll (1957).

241 developed by the psychologist Gordon Gallup Jr.: Gallup (1970).

241 some great apes: As reported in a paper with one of my all-time favorite titles: "Another gorilla, *Gorilla gorilla gorilla*, recognizes himself in the mirror" (Posada & Colell, 2007).

242 no convincing evidence: For a review, see Gallup & Anderson (2020). The cleaner wrasse debate is recounted in Kohda et al. (2019) and de Waal (2019).

242 "dognition": Gallup & Anderson (2018).

243 monkeys have even been trained: Cowey & Stoerig (1995).

243 a primate equivalent: Boly et al. (2013).

243 transplanted there in 1938: Kessler & Rawlins (2016).

244 **In one video:** www.ted.com/talks/frans_de_waal_do_animals_have_morals.

244 **failed the mirror test:** One study found that rhesus monkeys can pass the mirror test after several weeks of being trained to do so (L. Chang et al., 2017). But passing the test after extensive training is very different from the spontaneous use of mirrors for self-recognition.

245 *A Catalog of Body Patterning in Cephalopoda*: Borrelli et al. (2006). Taking this wonderful but extremely heavy book home totally broke my Ryanair baggage allowance.

246 **mind of an alien:** In 2015, I was invited to write a book chapter on "alien consciousness." After fretting for some time, I decided to write about actually existing octopuses instead of speculating about potentially existing aliens (Seth, 2016a). The parallel between octopuses and aliens has also been explored in Denis Villeneuve's artful 2016 film *Arrival*.

246 **"If we want to understand *other* minds":** Godfrey-Smith (2017), p. 10.

247 **octopus brains lack myelin:** Hochner (2012); Shigeno et al. (2018).

247 **Octopus consciousness:** Carls-Diamante (2017).

248 **Some researchers have suggested:** Liscovitch-Brauer et al. (2017).

248 **their cognitive abilities:** Fiorito & Scotto (1992). See also D. B. Edelman & Seth (2009); Mather (2019); and, for a classic text on cephalopod behavior, Hanlon & Messenger (1996).

248 **octopus caught out in the open:** www.bbc.co.uk/programs/p05nzfn1. There is similarly impressive footage in the 2020 documentary *My Octopus Teacher* by Pippa Ehrlich and James Reed, which features a filmmaker—Craig Foster—who develops a surprisingly close relationship with an octopus.

248 **meld into the background:** The marine biologist Roger Hanlon has captured many examples of octopus camouflage on video. Here's one of the best: www.youtube.com/watch?v=JSq8nghQZqA.

249 **central brain might not even know:** Mather (2019); Messenger (2001).

250 **octopuses can taste with their suckers:** van Giesen et al. (2020).

251 **macabre experiments:** Nesher et al. (2014).

252 **animals for which the "lights are on":** See D. B. Edelman & Seth (2009) and D. B. Edelman et al. (2005) for more on nonmammalian consciousness.

252 **Many bird species:** Clayton et al. (2007); Jao Keehn et al. (2019); Pepperberg & Gordon (2005); Pepperberg & Shive (2001).

252 **likely have conscious experiences:** Intriguingly, while bird brains have something similar to mammalian cortex (called the "pallium"), they lack any equivalent to the corpus callosum, which connects the two cortical hemispheres in mammals. Birds might therefore represent a kind of "natural split brain," raising questions about the unity of avian consciousness (Xiao & Gunturkun, 2009). Noah Strycker's *The Thing with Feathers* (2014) is a lovely introduction to avian cognition and behavior.

252 **As we move further out:** For excellent treatments of the evolution of consciousness, see Feinberg & Mallatt (2017); Ginsburg & Jablonka (2019); LeDoux (2019).

252 **Decisions about animal welfare:** See Birch (2017) for a useful overview, motivated by the argument that we should give animals the "benefit of the doubt" when evidence is inconclusive (an argument known more formally as the "precautionary principle"). Notably, the European Union decided in 2010 to include cephalopods under animal welfare legislation (directive 2010/63/EU).

253 **insect brains do possess:** Entler et al. (2016).

253 **fruit fly:** Khuong et al. (2019).

253 effective across *all* animals: Kelz & Mashour (2019).

254 possible conscious minds: Jonathan Birch, Alexandra Schnell, and Nicola Clayton usefully suggest the term "consciousness profile" for characterizing interspecies variation in conscious experience. They propose five dimensions of variation: perceptual richness, evaluative richness, unity, temporality, and selfhood. See Birch et al. (2020).

13. MACHINE MINDS

255 a golem: In his 1964 book *God and Golem Inc.*, the polymathic pioneer Norbert Wiener treated golems as central to his speculations about the risks of future AI.

256 vast mound of paper clips: In the parable of the paper clip maximizer, an AI is designed to make as many paper clips as possible. Because this AI lacks human values but is otherwise very smart, it destroys the world in its successful attempt to do so. See Bostrom (2014).

258 so-called "Singularity" hypothesis: See Shanahan (2015) for a refreshingly sober take on the Singularity hypothesis.

259 and intelligence can exist without consciousness: It is tempting to say that consciousness and intelligence are doubly dissociable—that each can exist without the other. But this would not be quite right. While I believe that intelligence can exist without consciousness, it could be that consciousness requires a nonzero level of intelligence.

259 not one single scale: The idea of multidimensional consciousness (and intelligence) recalls Jonathan Birch and colleagues' concept of a "consciousness profile" (Birch et al., 2020), as well as the multidimensional approach to levels of consciousness in humans, proposed by Tim Bayne, Jakob Hohwy, and Adrian Owen (Bayne et al., 2016).

261 The authors equivocate: Dehaene et al. (2017). Global availability corresponds to the popular global workspace theory of consciousness, whereas "self-monitoring" captures aspects of higher-order theories. We briefly met both of these theories in chapter 1. The authors of the *Science* paper explicitly admit that they may be leaving out the "experiential" component of consciousness. For me, this is leaving out too much.

262 machines might appear: This possibility arises because IIT accepts one part of functionalism (substrate independence) but not the other (sufficiency of input-output mappings). Some mechanisms, notably feedforward artificial neural networks of adequate size, can implement arbitrarily complex input-output mappings. These mechanisms, implemented the right way, might give the outward appearance of intelligence and/or consciousness. But purely feedforward networks generate no integrated information at all—some recurrency or "loopiness" is always needed. IIT therefore licenses the concept of a "behavioral zombie," which, as I explained in chapter 1, is an artifact that appears conscious from the outside, but has no consciousness. See Tononi & Koch (2015).

263 continually regenerating the conditions: These ideas relate closely to the concept of autopoiesis (pronounced "auto-poi-ee-sis," from the Greek words for "self" and "creation"), developed by Chilean biologist Humberto Maturana. An autopoietic system is one that is capable of maintaining and reproducing itself, which includes producing the physical components needed for its continued existence as a system. Although autopoeisis is first and foremost a theory of the cell, there are intriguing links between cellular autopoiesis and the free energy principle (see chapter 10). Both suggest a strong continuity between "life" and "mind," which in turn suggests that there is more to mind (and also to consciousness) than simply what a system "does" (Kirchhoff, 2018; Maturana & Varela,

1980). I was lucky enough to meet Maturana—who died in May 2021 at the age of 92—in January 2019, in his home city of Santiago, where we spent time sipping coffee and discussing these ideas in a shady café garden in the Barrio Providencia.

264 **Turing test, the famous yardstick:** In Alan Turing's original "imitation game," there are two humans of the same gender and a machine. The machine and one human—the collaborator—are both pretending to be a human of the opposite gender. The other human has to decide which is the machine and which is the collaborator (Turing, 1950).

265 **"the Garland test":** This term was coined by Murray Shanahan, whose book *Embodiment and the Inner Life* (2010) was one of the inspirations behind *Ex Machina*.

265 **noisy proclamations:** www.reading.ac.uk/news-archive/press-releases/pr583836.html.

265 **When the chatbot won:** "Eugene Goostman is a real boy—the Turing test says so." *Guardian* Pass notes, 9 June 2014. See https://www.theguardian.com/technology/short cuts/2014/jun/09/eugene-goostman-turing-test-computer-program.

265 **the humans failed:** The description of the Turing test as a test of "human gullibility" comes from a 2015 *New York Times* article by John Markoff. "Software is smart enough for SAT, but still far from intelligent" *New York Times* (21 September 2015). See www .nytimes.com/2015/09/21/technology/personaltech/software-is-smart-enough-for-sat -but-still-far-from-intelligent.html.

266 **vast artificial neural network:** GPT stands for "Generative Pre-trained Transformer"—a type of neural network specialized for language prediction and generation. These networks are trained using an unsupervised deep learning approach essentially to "predict the next word" given a previous word or text snippet. GPT-3 has an astonishing 175 billion parameters and was trained on some 45 terabytes of text data. See https://openai .com/blog/openai-api/; and for technical details: https://arxiv.org/abs/2005:14165.

266 **it does not understand:** Of course this depends on what is meant by "understanding." Some might say that human "understanding" is no different in kind from the sort of "understanding" displayed by GPT-3. The cognitive scientist Gary Marcus argues against this position, and I agree with him. See www.technologyreview.com/2020/08/22/1007 539/gpt3-openai-language-generator-artificial-intelligence-ai-opinion/.

266 **a five-hundred-word essay:** "A robot wrote this entire article. Are you scared yet, human?" *Guardian* Opinion, 8 September 2020. See www.theguardian.com/commentisfree/2020 /sep/08/robot-wrote-this-article-gpt-3. It is unclear how representative this example is.

267 **the most common feeling:** Becker-Asano et al. (2010).

268 **theories about why the uncanny valley exists:** Mori et al. (2012).

269 **"deepfake" technologies:** To "deepfake" is to generate a realistic but fake video, usually of a human face, using machine learning to combine a source and a target video. In a widely disseminated example from 2017, the deepfake method was used to create convincing videos of Barack Obama saying things that he did not say (www.youtube.com/watch?v= cQ54GDm1eL0). A series of TikTok videos deepfaking Tom Cruise, released in 2021, raises the bar substantially (https://www.theverge.com/22303756/tiktok-tom-cruise -impersonator-deepfake).

270 **vast uncontrolled global experiment:** The AI researcher Stuart Russell eloquently describes the threats posed by current and near-future AI, as well as ways to redesign AI systems to avoid them, in his book *Human Compatible* (2019). Nina Schick does a similarly excellent job for the threat posed by deepfakes (Schick, 2020).

271 **"intelligent tools, not colleagues"**: "Philosopher Daniel Dennett on AI, robots and religion," *Financial Times*, 3 March 2017. See https://www.ft.com/content/96187a7a-fce5 -11e6-96f8-3700c5664d30.

272 **immediate thirty-year moratorium**: Metzinger (2021).

273 **creating new forms of life**: Emmanuelle Charpentier and Jennifer Doudna won the 2020 Nobel Prize in Chemistry for their contributions to developing the CRISPR technique. The synthetic *E. coli* was created in the laboratory of Jason Chin; see Fredens et al. (2019).

273 **coordinated waves of electrical activity**: Trujillo et al. (2019).

274 **highly unlikely to be conscious**: I investigated the possibility of organoid consciousness in a recent paper with Tim Bayne and Marcello Massimini (Bayne et al., 2020).

274 **ethical urgency**: These issues are being taken seriously. In the summer of 2020, I—along with several other neuroscientists—was invited to speak at a US National Academy joint committee convened to help establish regulatory and legal frameworks for research involving both organoids and chimeras (animals genetically modified to express specific human characteristics). See www.nationalacademies.org/our-work/ethical-legal-and-reg ulatory-issues-associated-with-neural-chimeras-and-organoids.

274 **"We want to make farms"**: Carl Zimmer, "Organoids are not brains. How are they making brain waves?" *New York Times*, 29 August 2019. See www.nytimes.com/2019/08/29 /science/organoids-brain-alysson-muotri.html.

274 **a favorite trope of futurists**: For a measured discussion of the prospects and pitfalls of mind uploading, see Schneider (2019).

275 **virtual sentient agents**: The simulation argument runs like this (Bostrom, 2003). A sufficiently far-future civilization that has avoided extinguishing itself would likely have access to vast computational resources. Some members of this civilization may be inclined to run detailed computer simulations of their ancestors. Given that a great many such simulations could be run, it would therefore be rational for any individual experiencing life now to conclude that they were more likely to be among the simulated minds than among the original biological humans. As Bostrom puts it: "If we don't think that we are currently living in a computer simulation, we are not entitled to believe that we will have descendants who will run lots of such simulations of their forebears" (Bostrom, 2003, p. 243). One among many issues I have with this argument is that it assumes that functionalism is true: that, when it comes to consciousness, simulation is equivalent to instantiation. As I've mentioned before, I don't think functionalism is a safe assumption.

EPILOGUE

277 **a living but isolated island of cortex**: Since it is still connected to a blood supply and "alive," could this disconnected hemisphere sustain its own isolated consciousness? Potential "islands of awareness" like this might also occur in other emerging neurotechnologies, such *ex cranio* reanimated pig brains and the cerebral organoids I described in the previous chapter. We discuss all these cases in Bayne et al. (2020).

278 **"It is widely agreed"**: Chalmers (1995b), p. 201.

282 **may be aspects of perception**: See also Hoffman (2019).

REFERENCES

Albright, T. D. (2012). "On the perception of probable things: neural substrates of associative memory, imagery, and perception." *Neuron*, 74(2), 227–45.

Allen, M., Frank, D., Schwarzkopf, D. S., et al. (2016). "Unexpected arousal modulates the influence of sensory noise on confidence." *Elife*, 5, e18103.

Anscombe, G. E. M. (1959). *An Introduction to Wittgenstein's Tractatus*. London: St. Augustine's Press.

Aru, J., Bachmann, T., Singer, W., et al. (2012). "Distilling the neural correlates of consciousness." *Neuroscience and Biobehavioral Reviews*, 36(2), 737–46.

Ashby, W. R. (1952). *Design for a Brain*. London: Chapman and Hall.

Ashby, W. R. (1956). *An Introduction to Cybernetics*. London: Chapman and Hall.

Aspell, J. E., Heydrich, L., Marillier, G., et al. (2013). "Turning the body and self inside out: Visualized heartbeats alter bodily self-consciousness and tactile perception." *Psychological Science*, 24(12), 2445–53.

Baars, B. J. (1988). *A Cognitive Theory of Consciousness*. New York, NY: Cambridge University Press.

Barber, T. X. (1961). "Physiological effects of 'hypnosis.'" *Psychological Bulletin*, 58, 390–419.

Barnes, J. (2008). *Nothing to Be Frightened Of*. New York, NY: Knopf.

Barnett, L., Muthukumaraswamy, S. D., Carhart-Harris, R. L., et al. (2020). "Decreased directed functional connectivity in the psychedelic state." *Neuroimage*, 209, 116462.

Barrett, A. B., & Seth, A. K. (2011). "Practical measures of integrated information for time-series data." *PLoS Computational Biology*, 7(1), e1001052.

Barrett, L. F. (2017a). *How Emotions Are Made: The Secret Life of the Brain*. Boston, MA: Houghton Mifflin Harcourt.

Barrett, L. F. (2017b). "The theory of constructed emotion: an active inference account of interoception and categorization." *Social Cognitive and Affective Neuroscience*, 12(1), 1–23.

Barrett, L. F., & Satpute, A. B. (2019). "Historical pitfalls and new directions in the neuroscience of emotion." *Neuroscience Letters*, 693, 9–18.

Barrett, L. F., & Simmons, W. K. (2015). "Interoceptive predictions in the brain." *Nature Reviews Neuroscience*, 16(7), 419–29.

Bauby, J. D. (1997). *The Diving Bell and the Butterfly*. Paris: Robert Laffont.

Bayne, T. (2008). "The phenomenology of agency." *Philosophy Compass*, 3(1), 182–202.

Bayne, T. (2010). *The Unity of Consciousness*. Oxford: Oxford University Press.

Bayne, T. (2018). "On the axiomatic foundations of the integrated information theory of consciousness," *Neuroscience of Consciousness*, 1, niy007.

Bayne, T., Hohwy, J., & Owen, A. M. (2016). "Are There Levels of Consciousness?" *Trends in Cognitive Sciences*, 20(6), 405–13.

Bayne, T., Seth, A. K., & Massimini, M. (2020). "Are there islands of awareness?" *Trends in Neurosciences*, 43(1), 6–16.

Bechtel, W., & Williamson, R. C. (1998). "Vitalism." In E. Craig (ed.), *Routledge Encyclopedia of Philosophy*. London: Routledge.

Becker-Asano, C., Ogawa, K., Nishio, S., et al. (2010). "Exploring the uncanny valley with Geminoid HI-1 in a real-world application." *IADIS International Conferences Interfaces and Human Computer Interaction*, 121–28.

Berger, J. (1972). *Ways of Seeing*. London: Penguin.

Birch, J. (2017). "Animal sentience and the precautionary principle." *Animal Sentience*, 16(1).

Birch, J., Schnell, A. K., & Clayton, N. S. (2020). "Dimensions of Animal Consciousness." *Trends in Cognitive Sciences*, 24(10), 789–801.

Blake, R., Brascamp, J., & Heeger, D. J. (2014). "Can binocular rivalry reveal neural correlates of consciousness?" *Philosophical Transactions of the Royal Society B: Biological Sciences*, 369(1641), 20130211.

Blanke, O., Landis, T., Spinelli, L., et al. (2004). "Out-of-body experience and autoscopy of neurological origin." *Brain*, 127 (Pt 2), 243–58.

Blanke, O., Slater, M., & Serino, A. (2015). "Behavioral, neural, and computational principles of bodily self-consciousness." *Neuron*, 88(1), 145–66.

Block, N. (2005). "Two neural correlates of consciousness." *Trends in Cognitive Sciences*, 9(2), 46–52.

Boly, M., Seth, A. K., Wilke, M., et al. (2013). "Consciousness in humans and non-human animals: recent advances and future directions." *Frontiers in Psychology*, 4, 625.

Borrelli, L., Gherardi, F., & Fiorito, G. (2006). *A Catalog of Body Patterning in Cephalopoda*. Florence: Firenze University Press.

Bostrom, N. (2003). "Are you living in a computer simulation?" *Philosophical Quarterly*, 53(11), 243–55.

Bostrom, N. (2014). *Superintelligence: Paths, Dangers, Strategies*. Oxford: Oxford University Press.

Botvinick, M., & Cohen, J. (1998). "Rubber hands 'feel' touch that eyes see." *Nature*, 391(6669), 756.

Brainard, D. H., & Hurlbert, A. C. (2015). "Color vision: understanding #TheDress." *Current Biology*, 25(13), R551–54.

Brass, M., & Haggard, P. (2007). "To do or not to do: the neural signature of self-control." *Journal of Neuroscience*, 27(34), 9141–45.

Brass, M., & Haggard, P. (2008). "The what, when, whether model of intentional action." *Neuroscientist*, 14(4), 319–25.

Braun, N., Debener, S., Spychala, N., et al. (2018). "The senses of agency and ownership: a review." *Frontiers in Psychology*, 9, 535.

Brembs, B. (2011). "Toward a scientific concept of free will as a biological trait: spontaneous actions and decision-making in invertebrates." *Proceedings of the Royal Society B: Biological Sciences*, 278(1707), 930–39.

Brembs, B. (2020). "The brain as a dynamically active organ." *Biochemical and Biophysical Research Communications*. doi: 10.1016/j.bbrc.2020.12.011

Brener, J., & Ring, C. (2016). "Towards a psychophysics of interoceptive processes: the measurement of heartbeat detection." *Philosophical Transactions of the Royal Society B: Biological Sciences*, 371(1708), 20160015.

Brown, H., Adams, R. A., Parees, I., et al. (2013). "Active inference, sensory attenuation and illusions." *Cognitive Processing*, 14(4), 411–27.

Brown, R., Lau, H., & LeDoux, J. E. (2019). "Understanding the higher-order approach to consciousness." *Trends in Cognitive Sciences*, 23(9), 754–68.

Brugger, P., & Lenggenhager, B. (2014). "The bodily self and its disorders: neurological, psychological and social aspects." *Current Opinion in Neurology*, 27(6), 644–52.

Bruineberg, J., Dolega, K., Dewhurst, J., et al. (2020). "The Emperor's new Markov blankets." http://philsci-archive.pitt.edu/18467.

Bruner, J. S., & Goodman, C. C. (1947). "Value and need as organizing factors in perception." *Journal of Abnormal and Social Psychology*, 42(1), 33–44.

Buckley, C., Kim, C. S., McGregor, S., and Seth, A. K. (2017). "The free energy principle for action and perception: a mathematical review." *Journal of Mathematical Psychology*, 81, 55–79.

Burns, J. M., & Swerdlow, R. H. (2003). "Right orbitofrontal tumor with pedophilia symptom and constructional apraxia sign." *Archives of Neurology*, 60(3), 437–40.

Buzsáki, G. (2019). *The Brain from Inside Out*. Oxford: Oxford University Press.

Byrne, A., & Hilbert, D. (2011). "Are colors secondary qualities?" In L. Nolan (ed.), *Primary and Secondary Qualities: The Historical and Ongoing Debate*. Oxford: Oxford University Press, 339–61.

Caramazza, A., Anzellotti, S., Strnad, L., et al. (2014). "Embodied cognition and mirror neurons: a critical assessment." *Annual Review of Neuroscience*, 37, 1–15.

Carhart-Harris, R. L., Erritzoe, D., Williams, T., et al. (2012). "Neural correlates of the psychedelic state as determined by fMRI studies with psilocybin." *Proceedings of the National Academy of Sciences of the USA*, 109(6), 2138–43.

Carls-Diamante, S. (2017). "The octopus and the unity of consciousness." *Biology and Philosophy*, 32, 1269–87.

Casali, A. G., Gosseries, O., Rosanova, M., et al. (2013). "A theoretically based index of consciousness independent of sensory processing and behavior." *Science Translational Medicine*, 5(198), 198ra105.

Casarotto, S., Comanducci, A., Rosanova, M., et al. (2016). "Stratification of unresponsive patients by an independently validated index of brain complexity." *Annals of Neurology*, 80(5), 718–29.

Caspar, E. A., Christensen, J. F., Cleeremans, A., et al. (2016). "Coercion changes the sense of agency in the human brain." *Current Biology*, 26(5), 585–92.

Chalmers, D. J. (1995a). "Facing up to the problem of consciousness." *Journal of Consciousness Studies*, 2(3), 200–219.

Chalmers, D. J. (1995b). "The puzzle of conscious experience." *Scientific American*, 273(6), 80–86.

Chalmers, D. J. (1996). *The Conscious Mind: In Search of a Fundamental Theory*. New York, NY: Oxford University Press.

Chalmers, D. J. (2018). "The meta-problem of consciousness." *Journal of Consciousness Studies*, 25(9–10), 6–61.

Chang, H. (2004). *Inventing Temperature: Measurement and Scientific Progress*. New York, NY: Oxford University Press.

Chang, L., Zhang, S., Poo, M. M., et al. (2017). "Spontaneous expression of mirror self-recognition in monkeys after learning precise visual-proprioceptive association for mirror images." *Proceedings of the National Academy of Sciences of the USA*, 114(12), 3258–63.

Churchland, P. S. (1996). "The hornswoggle problem." *Journal of Consciousness Studies*, 3(5–6), 402–8.

Cisek, P. (2007). "Cortical mechanisms of action selection: the affordance competition hypothesis." *Philosophical Transactions of the Royal Society B: Biological Sciences*, 362(1485), 1585–99.

Clark, A. (2013). "Whatever next? Predictive brains, situated agents, and the future of cognitive science." *Behavioral and Brain Sciences*, 36(3), 181–204.

Clark, A. (2016). *Surfing Uncertainty*. Oxford: Oxford University Press.

Clayton, N. S., Dally, J. M., & Emery, N. J. (2007). "Social cognition by food-caching corvids. The western scrub-jay as a natural psychologist." *Philosophical Transactions of the Royal Society B: Biological Sciences*, 362(1480), 507–22.

Cobb, M. (2020). *The Idea of the Brain: A History*. London: Profile Books.

Collier, R. (2012). "Hospital-induced delirium hits hard." *Canadian Medical Association Journal*, 184(1), 23–24.

Conant, R., & Ashby, W. R. (1970). "Every good regulator of a system must be a model of that system." *International Journal of Systems Science*, 1(2), 89–97.

Cotard, J. (1880). "Du délire hypocondriaque dans une forme grave de la mélancolie anxieuse. Mémoire lu à la Société médico-psychophysiologique dans la séance du 28 Juin 1880." *Annales Medico-Psychologiques*, 168–74.

Cowey, A., & Stoerig, P. (1995). "Blindsight in monkeys." *Nature*, 373(6511), 247–49.

Craig, A. D. (2002). "How do you feel? Interoception: the sense of the physiological condition of the body." *Nature Reviews Neuroscience*, 3(8), 655–66.

Craig, A. D. (2009). "How do you feel—now? The anterior insula and human awareness." *Nature Reviews Neuroscience*, 10(1), 59–70.

Craver, C., & Tabery, J. (2017). "Mechanisms in science." In *The Stanford Encyclopedia of Philosophy*. plato.stanford.edu/entries/science-mechanisms.

Crick, F., & Koch, C. (1990). "Toward a neurobiological theory of consciousness." *Seminars in the Neurosciences*, 2, 263–75.

Critchley, H. D., & Harrison, N. A. (2013). "Visceral influences on brain and behavior." *Neuron*, 77(4), 624–38.

Csikszentmihalyi, M. (1990). *Flow: The Psychology of Optimal Experience*. New York, NY: Harper & Row.

Damasio, A. (1994). *Descartes' Error*. London: Macmillan.

Damasio, A. (2000). *The Feeling of What Happens: Body and Emotion in the Making of Consciousness*: Harvest Books.

Damasio, A. (2010). *Self Comes to Mind: Constructing the Conscious Brain*. London: William Heinemann.

Darwin, C. (1872). *The Expression of Emotions in Man and Animals*. London: Fontana Press.

Davis, D. H., Muniz-Terrera, G., Keage, H. A., et al. (2017). "Association of delirium with cognitive decline in late life: a neuropathologic study of three population-based cohort studies." *JAMA Psychiatry*, 74(3), 244–51.

de Graaf, T. A., Hsieh, P. J., & Sack, A. T. (2012). "The 'correlates' in neural correlates of consciousness." *Neuroscience and Biobehavioral Reviews*, 36(1), 191–97.

de Haan, E. H., Pinto, Y., Corballis, P. M., et al. (2020). "Split-brain: what we know about cutting the corpus callosum now and why this is important for understanding consciousness." *Neuropsychological Review*, 30, 224–33.

de Lange, F. P., Heilbron, M., & Kok, P. (2018). "How do expectations shape perception?" *Trends in Cognitive Sciences*, 22(9), 764–79.

de Waal, F. B. M. (2019). "Fish, mirrors, and a gradualist perspective on self-awareness." *PLoS Biology*, 17(2), e3000112.

Dehaene, S., & Changeux, J. P. (2011). "Experimental and theoretical approaches to conscious processing." *Neuron*, 70(2), 200–227.

Dehaene, S., Lau, H., & Kouider, S. (2017). "What is consciousness, and could machines have it?" *Science*, 358(6362), 486–92.

Deisseroth, K. (2015). "Optogenetics: ten years of microbial opsins in neuroscience." *Nature Neuroscience*, 18(9), 1213–25.

Della Sala, S., Marchetti, C., & Spinnler, H. (1991). "Right-sided anarchic (alien) hand: a longitudinal study." *Neuropsychologia*, 29(11), 1113–27.

Demertzi, A., Tagliazucchi, E., Dehaene, S., et al. (2019). "Human consciousness is supported by dynamic complex patterns of brain signal coordination." *Science Advances*, 5(2), eaat7603.

Dennett, D. C. (1984). *Elbow Room: The Varieties of Free Will Worth Wanting*. Cambridge, MA: MIT Press.

Dennett, D. C. (1991). *Consciousness Explained*. Boston, MA: Little, Brown.

Dennett, D. C. (1998). "The myth of double transduction." In S. Hameroff, A. W. Kasniak, & A. C. Scott (eds.), *Toward a Science of Consciousness II: The Second Tucson Discussions and Debates*. Cambridge, MA: MIT Press, 97–101.

Dennett, D. C. (2003). *Freedom Evolves*. New York, NY: Penguin Books.

Dennett, D. C. (2015). "Why and how does consciousness seem the way it seems?" In T. Metzinger & J. M. Windt (eds.), *Open MIND*. Frankfurt-am-Main: MIND Group.

Dennett, D. C., & Caruso, G. (2021). *Just Deserts: Debating Free Will*. Cambridge, MA: Polity.

Deutsch, D. (2012). *The Beginning of Infinity: Explanations That Transform the World*. New York, NY: Penguin Books.

DiNuzzo, M., & Nedergaard, M. (2017). "Brain energetics during the sleep-wake cycle." *Current Opinion in Neurobiology*, 47, 65–72.

Donne, J. (1839). "Devotions upon emergent occasions: Meditation XVII" [1624]. In H. Alford (ed.), *The Works of John Donne*. Vol. 3. London: Henry Parker, 574–75.

Duffy, S. W., Vulkan, D., Cuckle, H., et al. (2020). "Effect of mammographic screening from age forty years on breast cancer mortality (UK Age trial): final results of a randomized, controlled trial." *Lancet Oncology*, 21(9), 1165–72.

Dupuy, J. P. (2009). *On the Origins of Cognitive Science: The Mechanization of Mind*. 2nd ed. Cambridge, MA: MIT Press.

Dutton, D. G., & Aron, A. P. (1974). "Some evidence for heightened sexual attraction under conditions of high anxiety." *Journal of Personal and Social Psychology*, 30(4), 510–17.

Edelman, D. B., Baars, B. J., & Seth, A. K. (2005). "Identifying hallmarks of consciousness in non-mammalian species." *Consciousness and Cognition*, 14(1), 169–87.

Edelman, D. B., & Seth, A. K. (2009). "Animal consciousness: a synthetic approach." *Trends in Neuroscience*, 32(9), 476–84.

Edelman, G. M. (1989). *The Remembered Present*. New York, NY: Basic Books.

Edelman, G. M., & Gally, J. (2001). "Degeneracy and complexity in biological systems." *Proceedings of the National Academy of Sciences of the USA*, 98(24), 13763–68.

Ehrsson, H. H. (2007). "The experimental induction of out-of-body experiences." *Science*, 317(5841), 1048.

Ekman, P. (1992). "An argument for basic emotions." *Cognition and Emotion*, 6(3-4), 169–200.

Entler, B. V., Cannon, J. T., & Seid, M. A. (2016). "Morphine addiction in ants: a new model for self-administration and neurochemical analysis." *Journal of Experimental Biology*, 219 (Pt 18), 2865–69.

Evans, E. P. (1906). *The Criminal Prosecution and Capital Punishment of Animals*. London: William Heinemann.

Feinberg, T. E., & Mallatt, J. M. (2017). *The Ancient Origins of Consciousness: How the Brain Created Experience*. Cambridge, MA: MIT Press.

Feldman, H., & Friston, K. J. (2010). "Attention, uncertainty, and free-energy." *Frontiers in Human Neuroscience*, 4, 215.

Felleman, D. J., & Van Essen, D. C. (1991). "Distributed hierarchical processing in the primate cerebral cortex." *Cerebral Cortex*, 1(1), 1–47.

Ferrarelli, F., Massimini, M., Sarasso, S., et al. (2010). "Breakdown in cortical effective connectivity during midazolam-induced loss of consciousness." *Proceedings of the National Academy of Sciences of the USA*, 107(6), 2681–86.

Fiorito, G., & Scotto, P. (1992). "Observational learning in *Octopus vulgaris*." *Science*, 256(5056), 545–47.

Firestone, C. (2013). "On the origin and status of the 'El Greco fallacy.'" *Perception*, 42(6), 672–74.

Fleming, S. M. (2020). "Awareness as inference in a higher-order state space." *Neuroscience of Consciousness*, 2020(1), niz020.

Fletcher, P. C., & Frith, C. D. (2009). "Perceiving is believing: a Bayesian approach to explaining the positive symptoms of schizophrenia." *Nature Reviews Neuroscience*, 10(1), 48–58.

Flounders, M. W., Gonzalez-Garcia, C., Hardstone, R., & He, B. J. (2019). "Neural dynamics of visual ambiguity resolution by perceptual prior." *Elife*, 8, e41861.

Formisano, R., D'Ippolito, M., Risetti, M., et al. (2011). "Vegetative state, minimally conscious state, akinetic mutism and Parkinsonism as a continuum of recovery from disorders of consciousness: an exploratory and preliminary study." *Functional Neurology*, 26(1), 15–24.

Frankish, K. (2017). *Illusionism as a Theory of Consciousness*. Exeter: Imprint Academic.

Frässle, S., Sommer, J., Jansen, A., et al. (2014). "Binocular rivalry: frontal activity relates to introspection and action but not to perception." *Journal of Neuroscience*, 34(5), 1738–47.

Fredens, J., Wang, K., de la Torre, D., et al. (2019). "Total synthesis of *Escherichia coli* with a recoded genome." *Nature*, 569(7757), 514–18.

Fried, I., Katz, A., McCarthy, G., et al. (1991). "Functional organization of human supplementary motor cortex studied by electrical stimulation." *Journal of Neuroscience*, 11(11), 3656–66.

Friston, K. J. (2009). "The free-energy principle: a rough guide to the brain?" *Trends in Cognitive Sciences*, 13(7), 293–301.

Friston, K. J. (2010). "The free-energy principle: a unified brain theory?" *Nature Reviews Neuroscience*, 11(2), 127–38.

Friston, K. J. (2018). "Am I self-conscious? (Or does self-organization entail self-consciousness?)." *Frontiers in Psychology*, 9, 579.

Friston, K. J., Daunizeau, J., Kilner, J., et al. (2010). "Action and behavior: a free-energy formulation." *Biological Cybernetics*, 102(3), 227–60.

Friston, K. J., Thornton, C., & Clark, A. (2012). Free-energy minimization and the dark-room problem. *Frontiers in Psychology*, 3, 130.

Frith, C. D. (2007). *Making Up the Mind: How the Brain Creates Our Mental World*. Oxford: Wiley-Blackwell.

Gallagher, S. (2008). "Direct perception in the intersubjective context." *Consciousness and Cognition*, 17(2), 535–43.

Gallese, V., Fadiga, L., Fogassi, L., et al. (1996). "Action recognition in the premotor cortex." *Brain*, 119 (Pt 2), 593–609.

Gallup, G. G. (1970). "Chimpanzees: self-recognition." *Science*, 167(86–87).

Gallup, G. G., & Anderson, J. R. (2018). "The 'olfactory mirror' and other recent attempts to demonstrate self-recognition in non-primate species." *Behavioral Processes*, 148, 16–19.

Gallup, G. G., & Anderson, J. R. (2020). "Self-recognition in animals: Where do we stand fifty years later? Lessons from cleaner wrasse and other species." *Psychology of Consciousness: Theory, Research, and Practice*, 7(1), 46–58.

Gasquet, J. (1991). *Cézanne: A Memoir with Conversations*. London: Thames & Hudson Ltd.

Gehrlach, D. A., Dolensek, N., Klein, A. S., et al. (2019). "Aversive state processing in the posterior insular cortex." *Nature Neuroscience*, 22(9), 1424–37.

Gibson, J. J. (1979). *The Ecological Approach to Visual Perception*. Hillsdale, NJ: Lawrence Erlbaum.

Gidon, A., Zolnik, T. A., Fidzinski, P., et al. (2020). "Dendritic action potentials and computation in human layer 2/3 cortical neurons." *Science*, 367(6473), 83–87.

Gifford, C., & Seth, A. K. (2013). *Eye Benders: The Science of Seeing and Believing*. London: Thames & Hudson.

Ginsburg, S., & Jablonka, E. (2019). *The Evolution of the Sensitive Soul: Learning and the Origins of Consciousness*. Cambridge, MA: MIT Press.

Godfrey-Smith, P. G. (1996). "Spencer and Dewey on life and mind." In M. Boden (ed.), *The Philosophy of Artificial Life*. Oxford: Oxford University Press, 314–31.

Godfrey-Smith, P. G. (2017). *Other Minds: The Octopus, the Sea, and the Deep Origins of Consciousness*. New York: Farrar, Straus and Giroux.

Goff, P. (2019). *Galileo's Error: Foundations for a New Science of Consciousness*. London: Rider.

Gombrich, E. H. (1961). *Art and Illusion: A Study in the Psychology of Pictorial Representation*. Ewing, NJ: Princeton University Press.

Goodstein, D. L. (1985). *States of Matter*. Chelmsford, MA: Courier Corporation.

Graziano, M. S. (2017). "The attention schema theory: A foundation for engineering artificial consciousness." *Frontiers in Robotics and AI*, 4, 60.

Gregory, R. L. (1980). "Perceptions as hypotheses." *Philosophical Transactions of the Royal Society B: Biological Sciences*, 290(1038), 181–97.

Grill-Spector, K., & Malach, R. (2004). "The human visual cortex." *Annual Review of Neuroscience*, 27, 649–77.

Haggard, P. (2008). "Human volition: toward a neuroscience of will." *Nature Reviews Neuroscience*, 9(12), 934–46.

Haggard, P. (2019). "The neurocognitive bases of human volition." *Annual Review of Psychology*, 70, 9–28.

Hanlon, J., & Messenger, J. B. (1996). *Cephalopod Behavior*. Cambridge: Cambridge University Press.

Harding, D. E. (1961). *On Having No Head*. London: The Shollond Trust.

Harris, S. (2012). *Free Will*. New York: Deckle Edge.

Harrison, N. A., Gray, M. A., Gianaros, P. J., et al. (2010). "The embodiment of emotional feelings in the brain." *Journal of Neuroscience*, 30(38), 12878–84.

Harvey, I. (2008). "Misrepresentations." In S. Bullock, J. Noble, R. Watson, & M. Bedau (eds.), *Artificial Life Xi: Proceedings of the 11th International Conference on the Simulation and Synthesis of Living Systems*. Cambridge, MA: MIT Press, 227–33.

Hatfield, G. (2002). *Descartes and the Meditations*. Abingdon: Routledge.

Haun, A. M. (2021). "What is visible across the visual field?" *Neuroscience of Consciousness*.

Haun, A. M., & Tononi, G. (2019). "Why does space feel the way it does? Toward a principled account of spatial experience." *Entropy*, 21(12), 1160.

He, K., Zhang, X., Ren, S., et al. (2016). "Deep residual learning for image recognition." *2016 IEEE Conference on Computer Vision and Pattern Recognition (CVPR)*.

Heilbron, M., Richter, D., Ekman, M., et al. (2020). "Word contexts enhance the neural representation of individual letters in early visual cortex." *Nature Communications*, 11(1), 321.

Herculano-Houzel, S. (2009). "The human brain in numbers: a linearly scaled-up primate brain." *Frontiers in Human Neuroscience*, 3, 31.

Herculano-Houzel, S. (2016). *The Human Advantage: A New Understanding of How Our Brain Became Remarkable*. Cambridge, MA: MIT Press.

Hochner, B. (2012). "An embodied view of octopus neurobiology." *Current Biology*, 22(20), R887–92.

Hoel, E. P., Albantakis, L., & Tononi, G. (2013). "Quantifying causal emergence shows that macro can beat micro." *Proceedings of the National Academy of Sciences of the USA*, 110(49), 19790–95.

Hoffman, D. (2019). *The Case Against Reality: Why Evolution Hid the Truth from Our Eyes*. London: W. W. Norton.

Hoffman, D., Singh, M., & Prakash, C. (2015). "The interface theory of perception." *Psychonomic Bulletin and Review*, 22, 1480–1506.

Hohwy, J. (2013). *The Predictive Mind*. Oxford: Oxford University Press.

Hohwy, J. (2014). "The self-evidencing brain." *Nous*, 50(2), 259–85.

Hohwy, J. (2020a). "New directions in predictive processing." *Mind and Language*, 35(2), 209–23.

Hohwy, J. (2020b). "Self-supervision, normativity and the free energy principle." *Synthese*. doi:10.1007/s11229-020-02622-2.

Hohwy, J., & Seth, A. K. (2020). "Predictive processing as a systematic basis for identifying the neural correlates of consciousness." *Philosophy and the Mind Sciences*, 1(2), 3.

Hurley, S., & Noë, A. (2003). "Neural plasticity and consciousness." *Biology and Philosophy*, 18, 131–68.

Husserl, E. (1960 [1931]). *Cartesian Meditations: An Introduction to Phenomenology*. The Hague: Nijhoff.

Inagaki, K., & Hatano, G. (2004). "Vitalistic causality in young children's naive biology." *Trends in Cognitive Sciences*, 8(8), 356–62.

James, W. (1884). "What is an emotion?" *Mind*, 9(34), 188–205.

James, W. (1890). *The Principles of Psychology*. New York: Henry Holt.

Jao Keehn, R. J., Iversen, J. R., Schulz, I., et al. (2019). "Spontaneity and diversity of movement to music are not uniquely human." *Current Biology*, 29(13), R621–R622.

Jensen, F. V. (2000). *Introduction to Bayesian Networks*. New York: Springer.

Jensen, M. P., Jamieson, G. A., Lutz, A., et al. (2017). "New directions in hypnosis research: strategies for advancing the cognitive and clinical neuroscience of hypnosis." *Neuroscience of Consciousness*, 3(1), nix004.

Kail, P. J. E. (2007). *Projection and Realism in Hume's Philosophy*. Oxford: Oxford University Press.

Kandel, E. R. (2012). *The Age of Insight: The Quest to Understand the Unconscious in Art, Mind, and Brain, from Vienna 1900 to the Present*. New York: Random House.

Kelz, M. B., & Mashour, G. A. (2019). "The biology of general anesthesia from paramecium to primate." *Current Biology*, 29(22), R1199–R1210.

Kessler, M. J., & Rawlins, R. G. (2016). "A seventy-five-year pictorial history of the Cayo Santiago rhesus monkey colony." *American Journal of Primatology*, 78(1), 6–43.

Khuong, T. M., Wang, Q. P., Manion, J., et al. (2019). "Nerve injury drives a heightened state of vigilance and neuropathic sensitization in *Drosophila*." *Science Advances*, 5(7), eaaw4099.

Kirchhoff, M. (2018). "Autopoeisis, free-energy, and the life-mind continuity thesis." *Synthese*, 195(6), 2519–40.

Kirchhoff, M., Parr, T., Palacios, E., et al. (2018). "The Markov blankets of life: autonomy, active inference and the free energy principle." *Journal of the Royal Society Interface*, 15(138), 20170792.

Koch, C. (2019). *The Feeling of Life Itself: Why Consciousness Is Widespread but Can't Be Computed*. Cambridge, MA: MIT Press.

Kohda, M., Hotta, T., Takeyama, T., et al. (2019). "If a fish can pass the mark test, what are the implications for consciousness and self-awareness testing in animals?" *PLoS Biology*, 17(2), e3000021.

Konkoly, K. R., Appel, K., Chabani, E., et al. (2021). "Real-time dialogue between experimenters and dreamers during REM sleep." *Current Biology*. 31(7), 1417–1427.

Kornhuber, H. H., & Deecke, L. (1965). "Changes in the brain potential in voluntary movements and passive movements in man: readiness potential and reafferent potentials." *Pflügers Archiv für die gesamte Physiologie des Menschen und der Tiere*, 284, 1–17.

Kuhn, G., Amlani, A. A., & Rensink, R. A. (2008). "Towards a science of magic." *Trends in Cognitive Sciences*, 12(9), 349–54.

Lakatos, I. (1978). *The Methodology of Scientific Research Programs: Philosophical Papers*. Cambridge: Cambridge University Press.

La Mettrie, J. O. de (1748). *L'Homme machine*. Leiden: Luzac.

Lau, H., & Rosenthal, D. (2011). "Empirical support for higher-order theories of conscious awareness." *Trends in Cognitive Sciences*, 15(8), 365–73.

LeDoux, J. (2012). "Rethinking the emotional brain." *Neuron*, 73(4), 653–76.

LeDoux, J. (2019). *The Deep History of Ourselves: The Four-Billion-Year Story of How We Got Conscious Brains*. New York, NY: Viking.

LeDoux, J., Michel, M., & Lau, H. (2020). "A little history goes a long way toward understanding why we study consciousness the way we do today." *Proceedings of the National Academy of Sciences of the USA*, 117(13), 6976–84.

Lemon, R. N., & Edgley, S. A. (2010). "Life without a cerebellum." *Brain*, 133 (Pt 3), 652–54.

Lenggenhager, B., Tadi, T., Metzinger, T., et al. (2007). "Video ergo sum: manipulating bodily self-consciousness." *Science*, 317(5841), 1096–99.

Lettvin, J. Y. (1976). "On seeing sidelong." *The Sciences*, 16, 10–20.

Libet, B. (1985). "Unconscious cerebral initiative and the role of conscious will in voluntary action." *Behavioral and Brain Sciences*, 8, 529–66.

Libet, B., Wright, E. W., Jr., & Gleason, C. A. (1983). "Preparation- or intention-to-act, in relation to pre-event potentials recorded at the vertex." *Electroencephalography and Clinical Neurophysiology*, 56(4), 367–72.

Lipton, P. (2004). *Inference to the Best Explanation*. Abingdon: Routledge.

Liscovitch-Brauer, N., Alon, S., Porath, H. T., et al. (2017). "Trade-off between transcriptome plasticity and genome evolution in cephalopods." *Cell*, 169(2), 191–202 e111.

Livneh, Y., Sugden, A. U., Madara, J. C., et al. (2020). "Estimation of current and future physiological states in insular cortex." *Neuron*, 105(6), 1094–1111.e10.

Luppi, A. I., Craig, M. M., Pappas, I., et al. (2019). "Consciousness-specific dynamic interactions of brain integration and functional diversity." *Nature Communications*, 10(1), 4616.

Lush, P. (2020). "Demand characteristics confound the rubber hand illusion." *Collabra Psychology*, 6, 22.

Lush, P., Botan, V., Scott, R. B., et al. (2020). "Trait phenomenological control predicts experience of mirror synaesthesia and the rubber hand illusion." *Nature Communications*, 11(1), 4853.

Lyamin, O. I., Kosenko, P. O., Korneva, S. M., et al. (2018). "Fur seals suppress REM sleep for very long periods without subsequent rebound." *Current Biology*, 28(12), 2000–2005 e2002.

Makari, G. (2016). *Soul Machine: The Invention of the Modern Mind*. London: W. W. Norton.

Marken, R. S., & Mansell, W. (2013). "Perceptual control as a unifying concept in psychology." *Review of General Psychology*, 17(2), 190–95.

Markov, N. T., Vezoli, J., Chameau, P., et al. (2014). "Anatomy of hierarchy: feedforward and feedback pathways in macaque visual cortex." *Journal of Comparative Neurology*, 522(1), 225–59.

Marr, D. (1982). *Vision: A Computational Investigation into the Human Representation and Processing of Visual Information*. New York: Freeman.

Mashour, G. A., Roelfsema, P., Changeux, J. P., et al. (2020). "Conscious processing and the global neuronal workspace hypothesis." *Neuron*, 105(5), 776–98.

Massimini, M., Ferrarelli, F., Huber, R., et al. (2005). "Breakdown of cortical effective connectivity during sleep." *Science*, 309(5744), 2228–32.

Mather, J. (2019). "What is in an octopus's mind?" *Animal Sentience*, 26(1), 1–29.

Maturana, H., & Varela, F. (1980). *Autopoiesis and Cognition: The Realization of the Living*. Dordrecht: D. Reidel.

McEwan, I. (2000). *Atonement*. New York: Anchor Books.

McGinn, C. (1989). "Can we solve the mind-body problem?" *Mind*, 98(391), 349–66.

McGrayne, S. B. (2012). *The Theory That Would Not Die: How Bayes' Rule Cracked the Enigma Code, Hunted Down Russian Submarines, and Emerged Triumphant from Two Centuries of Controversy*. New Haven, CT: Yale University Press.

McLeod, P., Reed, N., & Dienes, Z. (2003). "Psychophysics: how fielders arrive in time to catch the ball." *Nature*, 426(6964), 244–45.

Mediano, P. A. M., Seth, A. K., & Barrett, A. B. (2019). "Measuring integrated information: comparison of candidate measures in theory and simulation." *Entropy*, 21(1), 17.

Mele, A. (2009). *Effective Intentions: The Power of Conscious Will*. New York: Oxford University Press.

Melloni, L., Schwiedrzik, C. M., Muller, N., et al. (2011). "Expectations change the signatures and timing of electrophysiological correlates of perceptual awareness." *Journal of Neuroscience*, 31(4), 1386–96.

Merker, B. (2007). "Consciousness without a cerebral cortex: a challenge for neuroscience and medicine." *Behavioral and Brain Sciences*, 30(1), 63–81; discussion 81–134.

Merleau-Ponty, M. (1962). *Phenomenology of Perception*. London: Routledge & Kegan Paul.

Merleau-Ponty, M. (1964). "Eye and mind." In J. E. Edie (ed.), *The Primacy of Perception*. Evanston, IL: Northwestern University Press, 159–90.

Messenger, J. B. (2001). "Cephalopod chromatophores: neurobiology and natural history." *Biological Reviews of the Cambridge Philosophical Society*, 76(4), 473–528.

Metzinger, T. (2003a). *Being No One*. Cambridge, MA: MIT Press.

Metzinger, T. (2003b). "Phenomenal transparency and cognitive self-reference." *Phenomenology and the Cognitive Sciences*, 2, 353–93.

Metzinger, T. (2021). "Artificial suffering: an argument for a global moratorium on synthetic phenomenology." *Journal of Artificial Intelligence and Consciousness*, 8(1), 1–24.

Monroe, R. (1971). *Journeys out of the Body*. London: Anchor Press.

Monti, M. M., Vanhaudenhuyse, A., Coleman, M. R., et al. (2010). "Willful modulation of brain activity in disorders of consciousness." *New England Journal of Medicine*, 362(7), 579–89.

Mori, M., MacDorman, K. F., & Kageki, N. (2012). "The Uncanny Valley." *IEEE Robotics & Automation Magazine*, 19(2), 98–100.

Myles, P. S., Leslie, K., McNeil, J., et al. (2004). "Bispectral index monitoring to prevent awareness during anesthesia: the B-Aware randomized controlled trial." *Lancet*, 363(9423), 1757–63.

Naci, L., Sinai, L., & Owen, A. M. (2017). "Detecting and interpreting conscious experiences in behaviorally non-responsive patients." *Neuroimage*, 145 (Pt B), 304–13.

Nagel, T. (1974). "What is it like to be a bat?" *Philosophical Review*, 83(4), 435–50.

Nasraway, S. S., Jr., Wu, E. C., Kelleher, R. M., et al. (2002). "How reliable is the Bispectral Index in critically ill patients? A prospective, comparative, single-blinded observer study." *Critical Care Medicine*, 30(7), 1483–87.

Nesher, N., Levy, G., Grasso, F. W., et al. (2014). "Self-recognition mechanism between skin and suckers prevents octopus arms from interfering with each other." *Current Biology*, 24(11), 1271–75.

Nin, A. (1961). *Seduction of the Minotaur*. Denver, CO: Swallow Press.

O'Regan, J. K. (2011). *Why Red Doesn't Sound Like a Bell: Understanding the Feel of Consciousness*. Oxford: Oxford University Press.

O'Regan, J. K., & Noë, A. (2001). "A sensorimotor account of vision and visual consciousness." *Behavioral and Brain Sciences*, 24(5), 939–73; discussion 973–1031.

Orne, M. T. (1962). "On the social psychology of the psychological experiment: with particular reference to demand characteristics and their implications." *American Psychologist*, 17, 776–83.

Owen, A. M. (2017). *Into the Gray Zone: A Neuroscientist Explores the Border Between Life and Death*. London: Faber & Faber.

Owen, A. M., Coleman, M. R., Boly, M., et al. (2006). "Detecting awareness in the vegetative state." *Science*, 313(5792), 1402.

Palmer, C. E., Davare, M., & Kilner, J. M. (2016). "Physiological and perceptual sensory attenuation have different underlying neurophysiological correlates." *Journal of Neuroscience*, 36(42), 10803–12.

Palmer, C. J., Seth, A. K., & Hohwy, J. (2015). "The felt presence of other minds: Predictive processing, counterfactual predictions, and mentalising in autism." *Consciousness and Cognition*, 36, 376–89.

Panksepp, J. (2004). *Affective Neuroscience: The Foundations of Human and Animal Emotions*. Oxford: Oxford University Press.

Panksepp, J. (2005). "Affective consciousness: core emotional feelings in animals and humans." *Consciousness and Cognition*, 14(1), 30–80.

Park, H. D., & Blanke, O. (2019). "Coupling inner and outer body for self-consciousness." *Trends in Cognitive Sciences*, 23(5), 377–88.

Park, H. D., & Tallon-Baudry, C. (2014). "The neural subjective frame: from bodily signals to perceptual consciousness." *Philosophical Transactions of the Royal Society B: Biological Sciences*, 369(1641), 20130208.

Parvizi, J., & Damasio, A. (2001). "Consciousness and the brainstem." *Cognition*, 79(1–2), 135–60.

Penrose, R. (1989). *The Emperor's New Mind*. Oxford: Oxford University Press.

Pepperberg, I. M., & Gordon, J. D. (2005). "Number comprehension by a gray parrot (*Psittacus erithacus*), including a zero-like concept." *Journal of Comparative Psychology*, 119(2), 197–209.

Pepperberg, I. M., & Shive, H. R. (2001). "Simultaneous development of vocal and physical object combinations by a gray parrot (*Psittacus erithacus*): bottle caps, lids, and labels." *Journal of Comparative Psychology*, 115(4), 376–84.

Petkova, V. I., & Ehrsson, H. H. (2008). "If I were you: perceptual illusion of body swapping." *PLoS One*, 3(12), e3832.

Petzschner, F. H., Weber, L. A., Wellstein, K. V., et al. (2019). "Focus of attention modulates the heartbeat evoked potential." *Neuroimage*, 186, 595–606.

Petzschner, F. H., Weber, L. A. E., Gard, T., et al. (2017). "Computational psychosomatics and computational psychiatry: toward a joint framework for differential diagnosis." *Biological Psychiatry*, 82(6), 421–30.

Phillips, M. L., Medford, N., Senior, C., et al. (2001). "Depersonalization disorder: thinking without feeling." *Psychiatry Research*, 108(3), 145–60.

Pinto, Y., van Gaal, S., de Lange, F. P., et al. (2015). "Expectations accelerate entry of visual stimuli into awareness." *Journal of Vision*, 15(8), 13.

Pollan, M. (2018). *How to Change Your Mind*. New York, NY: Penguin.

Portin, P. (2009). "The elusive concept of the gene." *Hereditas*, 146(3), 112–17.

Posada, S., & Colell, M. (2007). "Another gorilla (*Gorilla gorilla gorilla*) recognizes himself in a mirror." *American Journal of Primatology*, 69(5), 576–83.

Powers, W. T. (1973). *Behavior: The Control of Perception*. Hawthorne, NY: Aldine de Gruyter.

Press, C., Kok, P., & Yon, D. (2020). "The perceptual prediction paradox." *Trends in Cognitive Sciences*, 24(1), 13–24.

Pressnitzer, D., Graves, J., Chambers, C., et al. (2018). "Auditory perception: Laurel and Yanny together at last." *Current Biology*, 28(13), R739–R741.

Raccah, O., Block, N., & Fox, K. (2021). "Does the prefrontal cortex play an essential role in consciousness? Insights from intracranial electrical stimulation of the human brain." *Journal of Neuroscience*, 41(10), 2076–87.

Rao, R. P., & Ballard, D. H. (1999). "Predictive coding in the visual cortex: a functional interpretation of some extra-classical receptive-field effects." *Nature Neuroscience*, 2(1), 79–87.

Reep, R. L., Finlay, B. L., & Darlington, R. B. (2007). "The limbic system in mammalian brain evolution." *Brain, Behavior and Evolution*, 70(1), 57–70.

Richards, B. A., Lillicrap, T. P., Beaudoin, P., et al. (2019). "A deep learning framework for neuroscience." *Nature Neuroscience*, 22(11), 1761–70.

Riemer, M., Trojan, J., Beauchamp, M., et al. (2019). "The rubber hand universe: on the impact of methodological differences in the rubber hand illusion." *Neuroscience and Biobehavioral Reviews*, 104, 268–80.

Rosas, F., Mediano, P. A. M., Jensen, H. J., et al. (2021). "Reconciling emergences: an information-theoretic approach to identify causal emergence in multivariate data." *PLoS Computational Biology*, 16(12), e1008289.

Roseboom, W., Fountas, Z., Nikiforou, K., et al. (2019). "Activity in perceptual classification networks as a basis for human subjective time perception." *Nature Communications*, 10(1), 267.

Rousseau, M. C., Baumstarck, K., Alessandrini, M., et al. (2015). "Quality of life in patients with locked-in syndrome: evolution over a six-year period." *Orphanet Journal of Rare Diseases*, 10, 88.

Russell, S. (2019). *Human Compatible: Artificial Intelligence and the Problem of Control*. New York, NY: Viking.

Sabra, A. I. (1989). *The Optics of Ibn Al-Haytham*. Books 1–3. London: The Warburg Institute.

Schachter, S., & Singer, J. E. (1962). "Cognitive, social, and physiological determinants of emotional state." *Psychological Review*, 69, 379–99.

Schartner, M. M., Carhart-Harris, R. L., Barrett, A. B., et al. (2017a). "Increased spontaneous MEG signal diversity for psychoactive doses of ketamine, LSD and psilocybin." *Scientific Reports*, 7, 46421.

Schartner, M. M., Pigorini, A., Gibbs, S. A., et al. (2017b). "Global and local complexity of intracranial EEG decreases during NREM sleep." *Neuroscience of Consciousness*, 3(1), niw022.

Schartner, M. M., Seth, A. K., Noirhomme, Q., et al. (2015). "Complexity of multi-dimensional spontaneous EEG decreases during propofol induced general anesthesia." *PLoS One*, 10(8), e0133532.

Schick, N. (2020). *Deepfakes and the Infocalypse: What You Urgently Need to Know*. Monterey, CA: Monoray.

Schneider, S. (2019). *Artificial You: AI and the Future of Your Mind*. Princeton, NJ: Princeton University Press.

Schurger, A., Sitt, J. D., & Dehaene, S. (2012). "An accumulator model for spontaneous neural activity prior to self-initiated movement." *Proceedings of the National Academy of Sciences of the USA*, 109(42), E2904–13.

Searle, J. (1980). "Minds, brains, and programs." *Behavioral and Brain Sciences*, 3(3), 417–57.

Seth, A. K. (2009). "Explanatory correlates of consciousness: theoretical and computational challenges." *Cognitive Computation*, 1(1), 50–63.

Seth, A. K. (2010). "Measuring autonomy and emergence via Granger causality." *Artificial Life*, 16(2), 179–96.

Seth, A. K. (2013). "Interoceptive inference, emotion, and the embodied self." *Trends in Cognitive Sciences*, 17(11), 565–73.

Seth, A. K. (2014a). "Darwin's neuroscientist: Gerald M. Edelman, 1929–2014." *Frontiers in Psychology*, 5, 896.

Seth, A. K. (2014b). "A predictive processing theory of sensorimotor contingencies: explaining the puzzle of perceptual presence and its absence in synaesthesia." *Cognitive Neuroscience*, 5(2), 97–118.

Seth, A. K. (2015a). "The cybernetic Bayesian brain: from interoceptive inference to sensorimotor contingencies." In J. M. Windt & T. Metzinger (eds.), *Open MIND*. Frankfurt am Main: MIND Group, 35(T). https://open-mind.net/papers/the-cybernetic-bayesian-brain.

Seth, A. K. (2015b). "Inference to the best prediction." In T. Metzinger & J. M. Windt (eds.), *Open MIND*. Frankfurt am Main: MIND Group, 35(R). https://open-mind.net/papers/inference-to-the-best-prediction.

Seth, A. K. (2016a). "Aliens on earth: what octopus minds can tell us about alien consciousness." In J. Al-Khalili (ed.), *Aliens*. London: Profile Books, 47–58.

Seth, A. K. (2016b). "The real problem." *Aeon*. aeon.co/essays/the-hard-problem-of-con sciousness-is-a-distraction-from-the-real-one.

Seth, A. K. (2017). "The fall and rise of consciousness science." In A. Haag (ed.), *The Return of Consciousness*, Riga: Ax:Son Johnson Foundation, 13–41.

Seth, A. K. (2018). "Consciousness: The last 50 years (and the next)." *Brain and Neuroscience Advances*, 2, 2398212818816019.

Seth, A. K. (2019a). "Being a beast machine: the origins of selfhood in control-oriented interoceptive inference." In M. Colombo, L. Irvine, & M. Stapleton (eds.), *Andy Clark and His Critics*. Oxford: Wiley-Blackwell, 238–54.

Seth, A. K. (2019b). "From unconscious inference to the Beholder's Share: predictive perception and human experience." *European Review*, 273(3), 378–410.

Seth, A. K. (2019c). "Our inner universes." *Scientific American*, 321(3), 40–47.

Seth, A. K., Baars, B. J., & Edelman, D. B. (2005). "Criteria for consciousness in humans and other mammals." *Consciousness and Cognition*, 14(1), 119–39.

Seth, A. K., Barrett, A. B., & Barnett, L. (2011a). "Causal density and integrated information as measures of conscious level." *Philosophical Transactions of the Royal Society A: Mathematical, Physical, and Engineering Sciences*, 369(1952), 3748–67.

Seth, A. K., Dienes, Z., Cleeremans, A., et al. (2008). "Measuring consciousness: relating behavioral and neurophysiological approaches." *Trends in Cognitive Sciences*, 12(8), 314–21.

Seth, A. K., & Friston, K. J. (2016). "Active interoceptive inference and the emotional brain." *Philosophical Transactions of the Royal Society B: Biological Sciences*, 371(1708), 20160007.

Seth, A. K., Izhikevich, E., Reeke, G. N., et al. (2006). "Theories and measures of consciousness: an extended framework." *Proceedings of the National Academy of Sciences of the USA*, 103(28), 10799–804.

Seth, A. K., Millidge, B., Buckley, C. L., et al. (2020). "Curious inferences: reply to Sun and Firestone on the dark room problem." *Trends in Cognitive Sciences*, 24(9), 681–83.

Seth, A. K., Suzuki, K., & Critchley, H. D. (2011b). "An interoceptive predictive coding model of conscious presence." *Frontiers in Psychology*, 2, 395.

Seth, A. K., Roseboom, W., Dienes, Z., & Lush, P. (2021). "What's up with the rubber hand illusion?" https://psyarxiv.com/b4qcy/.

Seth, A. K., & Tsakiris, M. (2018). "Being a beast machine: the somatic basis of selfhood." *Trends in Cognitive Sciences*, 22(11), 969–81.

Shanahan, M. P. (2010). *Embodiment and the Inner Life: Cognition and Consciousness in the Space of Possible Minds*. Oxford: Oxford University Press.

Shanahan, M. P. (2015). *The Technological Singularity*. Cambridge, MA: MIT Press.

Sherman, M. T., Fountas, Z., Seth, A. K., et al. (2020). "Accumulation of salient events in sensory cortex activity predicts subjective time." www.biorxiv.org/content/10:1101/2020:01.09:900423v4.

Shigeno, S., Andrews, P. L. R., Ponte, G., et al. (2018). "Cephalopod brains: an overview of current knowledge to facilitate comparison with vertebrates." *Frontiers in Physiology*, 9, 952.

Shugg, W. (1968). "The cartesian beast-machine in English literature (1663–1750)." *Journal of the History of Ideas*, 29(2), 279–92.

Silver, D., Schrittwieser, J., Simonyan, K., et al. (2017). "Mastering the game of Go without human knowledge." *Nature*, 550(7676), 354–59.

Simons, D. J., & Chabris, C. F. (1999). "Gorillas in our midst: sustained inattentional blindness for dynamic events." *Perception*, 28(9), 1059–74.

Solms, M. (2018). "The hard problem of consciousness and the free energy principle." *Frontiers in Physiology*, 9, 2714.

Solms, M. (2021). *The Hidden Spring: A Journey to the Source of Consciousness*. London: Profile Books.

Stein, B. E., & Meredith, M. A. (1993). *The Merging of the Senses*. Cambridge, MA: MIT Press.

Steiner, A. P., & Redish, A. D. (2014). "Behavioral and neurophysiological correlates of regret in rat decision-making on a neuroeconomic task." *Nature Neuroscience*, 17(7), 995–1002.

Sterling, P. (2012). "Allostasis: a model of predictive regulation." *Physiology and Behavior*, 106(1), 5–15.

Stetson, C., Fiesta, M. P., & Eagleman, D. M. (2007). "Does time really slow down during a frightening event?" *PLoS One*, 2(12), e1295.

Stoelb, B. L., Molton, I. R., Jensen, M. P., et al. (2009). "The efficacy of hypnotic analgesia in adults: a review of the literature." *Contemporary Hypnosis*, 26(1), 24–39.

Stoljar, D. (2017). "Physicalism." In E. N. Zalta (ed.), *The Stanford Encyclopedia of Philosophy* (Winter 2017 ed.). plato.stanford.edu/archives/win2017/entries/physicalism/.

Strawson, G. (2008). *Real Materialism and Other Essays*. Oxford: Oxford University Press.

Strycker, N. (2014). *The Thing with Feathers: The Surprising Lives of Birds and What They Reveal about Being Human*. New York, NY: Riverhead Books.

Suárez-Pinilla, M., Nikiforou, K., Fountas, Z., et al. (2019). "Perceptual content, not physiological signals, determines perceived duration when viewing dynamic, natural scenes." *Collabra Psychology*, 5(1), 55.

Sun, Z., & Firestone, C. (2020). "The dark room problem." *Trends in Cognitive Sciences*, 24(5), 346–48.

Sutherland, S. (1989). *International Dictionary of Psychology*. New York: Crossroad Classic.

Suzuki, K., Garfinkel, S. N., Critchley, H. D., & Seth, A. K. (2013). "Multisensory integration across exteroceptive and interoceptive domains modulates self-experience in the rubber-hand illusion." *Neuropsychologia*, 51(13), 2909–17.

Suzuki, K., Roseboom, W., Schwartzman, D. J., and Seth, A. K. (2017). "A deep-dream virtual reality platform for studying altered perceptual phenomenology." *Scientific Reports*, 7(1), 15982.

Suzuki, K., Schwartzman, D. J., Augusto, R., and Seth, A. K. (2019). "Sensorimotor contingency modulates breakthrough of virtual 3D objects during a breaking continuous flash suppression paradigm." *Cognition*, 187, 95–107.

Suzuki, K., Wakisaka, S., & Fujii, N. (2012). "Substitutional reality system: a novel experimental platform for experiencing alternative reality." *Scientific Reports*, 2, 459.

Swanson, L. R. (2016). "The predictive processing paradigm has roots in Kant." *Frontiers in Systems Neuroscience*, 10, 79.

Teasdale, G. M., & Murray, L. (2000). "Revisiting the Glasgow Coma Scale and Coma Score." *Intensive Care Medicine*, 26(2), 153–54.

Teufel, C., & Fletcher, P. C. (2020). "Forms of prediction in the nervous system." *Nature Reviews Neuroscience*, 21(4), 231–42.

Thompson, E. (2007). *Mind in Life: Biology, Phenomenology, and the Sciences of Mind*. Cambridge, MA: Harvard University Press.

Thompson, E. (2014). *Waking, Dreaming, Being: Self and Consciousness in Neuroscience, Meditation, and Philosophy*. New York, NY: Columbia University Press.

Timmermann, C., Roseman, L., Schartner, M., et al. (2019). "Neural correlates of the DMT experience assessed with multivariate EEG." *Scientific Reports*, 9(1), 16324.

Tong, F. (2003). "Out-of-body experiences: from Penfield to present." *Trends in Cognitive Sciences*, 7(3), 104–6.

Tononi, G. (2008). "Consciousness as integrated information: a provisional manifesto." *Biological Bulletin*, 215(3), 216–42.

Tononi, G. (2012). "Integrated information theory of consciousness: an updated account." *Archives italiennes de biologie*, 150(4), 293–329.

Tononi, G., Boly, M., Massimini, M., et al. (2016). "Integrated information theory: from consciousness to its physical substrate." *Nature Reviews Neuroscience*, 17(7), 450–61.

Tononi, G., & Edelman, G. M. (1998). "Consciousness and complexity." *Science*, 282(5395), 1846–51.

Tononi, G., & Koch, C. (2015). "Consciousness: here, there and everywhere?" *Philosophical Transactions of the Royal Society B: Biological Sciences*, 370(1668).

Tononi, G., Sporns, O., & Edelman, G. M. (1994). "A measure for brain complexity: relating functional segregation and integration in the nervous system." *Proceedings of the National Academy of Sciences of the USA*, 91(11), 5033–37.

Trujillo, C. A., Gao, R., Negraes, P. D., et al. (2019). "Complex oscillatory waves emerging from cortical organoids model early human brain network development." *Cell Stem Cell*, 25(4), 558–69 e557.

Tschantz, A., Barca, L., Maisto, D., et al. (2021). "Simulating homeostatic, allostatic and goal-directed forms of interoceptive control using active inference." https://www.biorxiv.org/content/10.1101/2021.02.16.431365v1.

Tschantz, A., Millidge, B., Seth, A. K., et al. (2020a). "Reinforcement learning through active inference." doi:https://arxiv.org/abs/2002:12636.

Tschantz, A., Seth, A. K., & Buckley, C. (2020b). "Learning action-oriented models." *PLoS Computational Biology*, 16(4), e1007805.

Tsuchiya, N., Wilke, M., Frässle, S., et al. (2015). "No-report paradigms: extracting the true neural correlates of consciousness." *Trends in Cognitive Sciences*, 19(12), 757–70.

Tulving, E. (1985). "Memory and consciousness." *Canadian Psychology*, 26, 1–12.

Turing, A. M. (1950). "Computing machinery and intelligence." *Mind*, 59, 433–60.

Uexküll, J. von (1957). "A stroll through the worlds of animals and men: a picture book of invisible worlds." In C. Schiller (ed.), *Instinctive Behavior: The Development of a Modern Concept*. New York: International Universities Press, 5.

van Giesen, L., Kilian, P. B., Allard, C. A. H., et al. (2020). "Molecular basis of chemotactile sensation in octopus." *Cell*, 183(3), 594–604 e514.

van Rijn, H., Gu, B. M., & Meck, W. H. (2014). "Dedicated clock/timing-circuit theories of time perception and timed performance." *Advances in Experimental Medicine and Biology*, 829, 75–99.

Varela, F. J. (1996). "Neurophenomenology: A methodological remedy for the hard problem." *Journal of Consciousness Studies*, 3, 330–50.

Varela, F. J., Thompson, E., & Rosch, E. (1993). *The Embodied Mind: Cognitive Science and Human Experience*. Cambridge, MA: MIT Press.

Walker, M. (2017). *Why We Sleep*. New York: Scribner.

Waller, B. (2011). *Against Moral Responsibility*. Cambridge, MA: MIT Press.

Wearing, D. (2005). *Forever Today: A Memoir of Love and Amnesia*. London: Corgi.

Wegner, D. (2002). *The Illusion of Conscious Will*. Cambridge, MA: MIT Press.

Weiser, T. G., Regenbogen, S. E., Thompson, K. D., et al. (2008). "An estimation of the global volume of surgery: a modelling strategy based on available data." *Lancet*, 372(9633), 139–44.

Wheeler, J. A. (1989). "Information, physics, quantum: the search for links." *Proceedings III International Symposium on Foundations of Quantum Mechanics*, Tokyo, 354–58.

Wiener, N. (1948). *Cybernetics: Or Control and Communication in the Animal and Machine*. Cambridge, MA: MIT Press.

Wiener, N. (1964). *God and Golem, Inc.* Cambridge, MA: MIT Press.

Williford, K., Bennequin, D., Friston, K., et al. (2018). "The projective consciousness model and phenomenal selfhood." *Frontiers in Psychology*, 9, 2571.

Winn, J., & Bishop, C. M. (2005). "Variational message passing." *Journal of Machine Learning Research*, 6, 661–94.

Wittmann, M. (2013). "The inner sense of time: how the brain creates a representation of duration." *Nature Reviews Neuroscience*, 14(3), 217–23.

Witzel, C., Racey, C., & O'Regan, J. K. (2017). "The most reasonable explanation of 'the dress': implicit assumptions about illumination." *Journal of Vision*, 17(2), 1.

Xiao, Q., & Gunturkun, O. (2009). "Natural split-brain? Lateralized memory for task contingencies in pigeons." *Neuroscience Letters*, 458(2), 75–78.

Zamariola, G., Maurage, P., Luminet, O., & Corneille, O. (2018). "Interoceptive accuracy scores from the heartbeat counting task are problematic: Evidence from simple bivariate correlations." *Biological Psychology*, 137, 12–17.

Zucker, M. (1945), *The Philosophy of American History*, vol. 1: *The Historical Field Theory*. New York: Arnold-Howard.

INDEX

ABOUT THE AUTHOR

Anil Seth is a professor of cognitive and computational neuroscience at the University of Sussex, and codirector of the Sackler Centre for Consciousness Science.